With a
Factor 8
to the
Mobility
System
of the Future

With a Factor 8 to the Mobility System of the Future

Cathy Macharis

Illustrations
Mathilde Guegan

stichting kunstboek

Content

FACTOR 8

Reducing CO₂
a factor of 8:

emissions by
Why?

"We're not going to make it to 2050" was the alarming headline of a 2022 article by the British economist, Umair Haque. Although pessimistic, this headline is fast becoming our reality. Much of Europe experienced a massive wave of heat this 2022 summer, and this is simply a prelude to what is yet to come. Australia battled floods just after recovering from wildfires in previous years, and Southeast Asia experienced up to 16 floods and four landslides in just the first half of the year. Simply put, several parts of the world have experienced one natural disaster after the other, and they are bracing for more. But let me put it in a clearer context.

I was on holiday in the French Alps, by the beautiful Lake Allos. A lake that is fed by glaciers whose water runs off via waterfalls. Every year, one can see how this process accelerates, how the level of the lake goes down and how the vegetation around the lake changes.

What am I getting at? Climate change is already here. We do not need to think that it is something abstract for the future. We are already experiencing it.

How did it come to this? Since the Industrial Revolution, many greenhouse gases have been released into the air, mainly by burning fossil fuels. This disturbs the natural greenhouse effect and results in climate change. In fact, you can picture the atmosphere as a shell around the earth, which we are now filling with CO_2 and other greenhouse gases such as methane and nitrous oxide, thickening the shell and forming a thick insulating layer. As a result, the heat from the sun that is returned by the earth as thermal radiation can no longer escape. The greenhouse gases, such as CO_2, methane, and nitrous oxide, absorb heat radiation in the atmosphere and subsequently radiate it in all directions, including back to the earth. This makes the earth warmer and causes climate change.

This greenhouse effect still had a positive effect until the Industrial Revolution. Without it, it would be freezing cold here. But since the Industrial Revolution, the volume of greenhouse gases that we pump into the atmosphere has increased enormously, so the insulation layer has become too thick. As a result, global temperature has already increased by 1.1

degrees[1], and if we continue at this rate, the temperature on earth will increase by 8.5 degrees - with the measures that are now being taken, by 3 to 4 degrees - which is an absolute disaster[2]. In fact, there is now a 50-50 possibility of hitting 1.5 degrees by 2026[3].

How does this happen? Trees and plants absorb CO_2 from the air as they grow. Since fossil fuels are created when tree and plant residues and other organic material are compressed into the earth's crust, the burning of fossil fuels such as gas, oil, and coal releases all the CO_2 that has been stored in them over the centuries. It is a process that happens under great pressure and takes millions of years. And that gas is released back into the atmosphere in large quantities during combustion. As for methane, it mainly comes from the decay process of plants.

The impact is clearly measurable: warming of the atmosphere and oceans, a change in the frequency and intensity of precipitation, a change in the activity of storms, a decrease in the amount of snow and ice, and an increase and acidification of the oceans.

This, in turn, has an impact on biodiversity, agriculture, health, the economy, and so on - in short, on the entire ecosystem. The figure below shows the relationship between the increasing level of CO_2 emissions in our atmosphere and temperature.

CO_2 is the main greenhouse gas that is emitted. It is responsible for two-thirds of global warming[4]. But there are also emissions of methane, nitrous oxide, and other greenhouse gases. The concentrations of these gases have been rising sharply over the last 150 years. Methane is one of the most vicious gases. It traps 80 times more heat than CO_2 and is responsible for 17% of the current warming. It does disappear from the atmosphere after nine years, but it then decomposes back into CO_2[5]. In the rest of the book, I will mainly talk about CO_2 emissions because they are the main contributor, but I will also mention other greenhouse gases when they are relevant[6]. Often, CO_2-equivalent is used. In this way, all gases are expressed with CO_2 as standard. To make this conversion, both the amount of heat that is retained and how long the gas stays in the atmosphere are considered. For example, the emission of 1kg of methane is equal to 25 CO_2-equivalents.

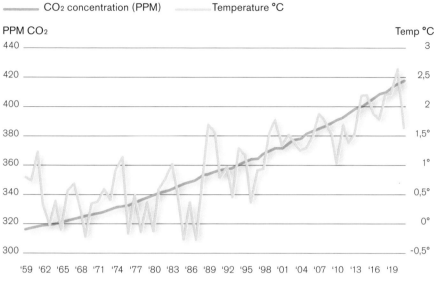

CO₂ concentration (PPM)   Temperature °C

PPM CO₂   Temp °C

Annual global temperature and CO_2 levels between 1959 and 2019[7].

The earth has already warmed up by over 1oC. Droughts, floods, forest fires, hurricanes… We see them on the news almost every day. But extremely cold periods are also part of climate change. These new weather conditions have a significant effect on the survival of some animal species, but also on our food supply. We may not be so aware of it anymore, but what we eat still has to be grown. In 2018, extreme weather conditions across Europe caused problems for agriculture[8]. Indeed, our food supply is at risk.

According to Al Gore, everyone has to have their aha-moment when it comes to climate. The moment when you suddenly realise, "damn, we really do have a problem". For me, that moment came a few years ago when I was travelling with my family to Senegal. Until then, the climate had always been a topic in my research, along with the many other effects of transport. But at that moment, I felt how climate change would impact the lives of the people in Africa. In Europe, we often imagine that life will go on as before, or at least we hope so. By the time there is another heat wave, we will have installed air conditioning. And when the weather is bad, we just turn up the heat. We often live very much sheltered from the weather. But when you live with nature, as you do in some parts of the world, you feel that small changes can indeed cause the harvest to fail and that such a loss cannot be compensated for by going to the supermarket to buy something extra. That realisation, my aha-moment, made me work on the theme even more. And indeed, the realisation that 'thinking the problem away' – as many do – will unfortunately not solve it. If we want to give our children as pleasant a living environment as we ourselves have had, we absolutely must tackle the problem and do so immediately.

If we continue like this, by the end of the century, the earth will be about 3 to 4°C warmer which means that several places will become inhabitable. Does this seem like a small change? It is not, as 3 to 8 degrees made the difference between our current climate and the last ice age, except this time it is the other way round. And if you don't care so much about cold or warm, what about mosquitoes that come bother you in winter as well? Or other tropical insects that are making their way north?

And climate change is very much about people. The many floods, storms and crop failures will cause people to flee from the climate effects and a new problem of climate refugees will arise. The war in Syria, for example, was partly caused by the great drought and subsequent crop failures.

All over, cities will suffer greatly from rising temperatures. Due to the rising sea level, large parts of Europe will be flooded[9]. By the end of this century, the sea level will have risen by two metres. This means that part of the land will be unliveable, and even worse some islands will disappear[10].

Not to mention the so-called tipping points: irreversible processes triggered by warming that will accelerate global warming. One example is the release of methane from beneath the ice caps. It will also take tens of thousands of years for nature to remove CO_2 emissions from the atmosphere. Trees are important for reducing emissions in part, and we must certainly work on that, but it is not enough. In 2018, it was reported that trees absorbed only about 21 billion tonnes of CO_2 of the 41 billion tonnes of CO_2 emitted[11]. The supposed "carbon capture and storage" technology cannot solve the rising emissions either. It is a system whereby CO_2 is extracted from the air and stored somewhere. The CO_2-filters are not only hugely expensive, but they also cannot cope with the large quantities. The entire planet would have to be filled with such extraction systems, and to remove 1 percent of annual emissions would require an investment of $400 billion in the technology. Putting everything under-ground would also involve additional costs[12]. So, it really comes down to reducing emissions.

During the climate negotiations in Paris, it was agreed that countries would do everything in their power to keep the further rise in temperature below 2°C, and preferably below 1.5°C. In order to achieve this, Europe has put forward the objective in its Climate and Energy Paper[13] of reducing greenhouse gases by at least 40% by 2030 in comparison with 1990 and by 80 to 98% by 2050 but this has now been revised to be at least 55% in 2030 and climate neutrality within the EU by 2050[14]. Consequently, in July 2021, a new series of legislative proposals were adopted to meet this newly set targets[15].

FACTOR 8

Climate and Transport

If we look at the sources of these emissions, the transport sector has a significant impact: in Europe, 32% of CO_2 emissions come from the transport sector[16]! Other important sectors that have an impact on the CO_2 emissions are the energy sector, buildings, agriculture, waste and industry.

Within the transport sector, road transport has the highest emissions (73.2%), with passenger transport accounting for 62.3% of emissions, vans for 11.9% and trucks for 25.8%. It is therefore vital to look at how we can reduce emissions in the road transport sector!

Aviation and maritime transport are also important sources of CO_2 emissions, and their importance is still growing fast. In the total European transport emissions, aviation is responsible for 13.9% And maritime transport for 12.9%. Moreover, these two sectors do not yet have a system with stringent regulations and clear targets.

For maritime transport, which is responsible for 2% of global emissions (as much as the total emissions of the whole of Belgium), there are yet no binding agreements for reducing the climate impact. Most ships still burn dirty heavy fuel oil, especially when they are in international waters. Unfortunately, this was not covered by the 2015 Paris Agreement, neither by the EU's Energy Taxation Directive. Although the IMO (International Maritime Organisation) has its own strategy to reduce emissions, these efficiency improvements are being counteracted by more and more transport, especially because globalisation and the growth of maritime transport go hand in hand. Hence, the European parliament is pushing to include maritime transport into the European Trading System (ETS, see box)[17].

Climate policy at European level is divided into two main groups of sectors: those that are covered by the Emissions Trading Scheme (EU ETS) and those, including transport, that are not. The EU Emissions Trading Scheme (EU ETS) is a mechanism to achieve emission reductions by setting a ceiling on emissions. Companies that succeed in reducing their emissions can sell their excess to other companies that have not yet succeeded in decarbonising and so must pay the price. An advantage of the system is also that the number of emission allowances can be slowly reduced, which means that the targets can be met, partly stimulated by the price mechanism. Sectors covered by the EU ETS are mainly aviation (continental flights within Europe only), the energy sector and energy-intensive industrial sectors (such as oil refineries and steel mills). ETS sectors account for about 40% of the EU's greenhouse gas emissions (EC, 2015; 2017). The non-ETS sectors mainly include transport, buildings, agriculture, waste and non-ETS industry, and cover about 60% of total EU domestic emissions (EC, 2018). By 2020, the non-ETS sector should reduce its emissions by 15% and by 35% CO_2 by 2030 compared to the 2005 baseline. Some experts, such as Jos Delbeke, argue that the transport sector should also be included in the ETS in order to allow the price mechanism to play its full role and, moreover, to reduce the number of emission allowances gradually. On the other hand, the mechanism is also much criticised because when there is too much supply of emission rights (as in the first period of the system), it makes the price of a tonne of CO_2 much too low. The EU has recently proposed an ETS2 to cover indeed road transport and heating, but this has sparked huge controversy so it is not clear if this will be implemented.

As earlier mentioned, aviation is responsible for 13.9% of European transport emissions[18]. If you look at it globally, not even 2.5% of global greenhouse gases. But of course, everything is only a small percentage. However, if aviation were a country, it would be the fifth largest polluter. Moreover, the sector is growing enormously, so are its emissions. Given that other sectors are reducing their emissions; aviation would be the sector responsible for 22% of global emissions in 2050[19]. As with maritime transport, this is a sector with a huge growth forecast that outstrips possible technological improvements. Aircraft are twice as efficient as in 1960[20] but compared to 1970 there are 14 times more passengers[21] (an increase of 5.5% per year[22]). The story here is very similar to that of maritime transport, and this sector too has remained mainly outside the scope of binding agreements. For flights within the European Union, European airlines have to buy emission rights. The price paid for this is about 1 euro per ticket and therefore has little effect on our purchasing behaviour[23]. The international aviation umbrella organisation ICAO has now set up a worldwide compensation system namely Corsia project in which international airlines purchase offset credits to compensate for their carbon emissions once these exceed the 2019 level.

It is also striking that the transport sector is the only sector that has not succeeded in reducing its CO_2 emissions below the 1990 level[24] while the other sectors, such as energy, industry, agriculture, households, etc., have managed to reduce their CO_2 emissions[25.]

In the figure below you can see that the transport sector is still on the rise: and international flights also continue to rise.

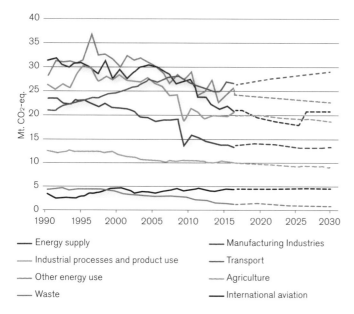

Total CO$_2$ emissions per sector, 1990-2017, projection of existing policies until 2030

Why is that? We have been talking about climate for years, yet we see no change. On the contrary, the root of this evolution is still the increasing demand for transport. Here, the explanation is simple and the same for the entire transport sector, so it extends to the aviation and maritime sector. Although there has been an improvement in terms of energy consumption per distance travelled, this technological progress is being cancelled out by an increase in demand. We will discuss this in more detail in the next section.

FACTOR 8

Underlying trends

Emissions of road transport not only depend on how much driving is done but also on the amount of energy is consumed during that driving. So, it is important to look at the origin of that energy. For example, is it petrol, diesel or electrical energy from renewable sources?

The volume of road transport in Europe has been increasing for most of the past 25 years (check figure below.) for both passenger transport (cars) and freight (trucks)[26]. Transport demand for freight roughly follows the growth of economic performance (expressed by the Gross Domestic Product [GDP]). This strong link with economic growth is very apparent in freight transport, which fell sharply during the 2008 crisis, for example. The connection of GDP growth to road passenger transport is less obvious, but the kilometres driven in cars on the roads of Europe also increased by 30% since 1995. The more people have jobs, but also the more people consume and produce, the more activity there is on the roads.

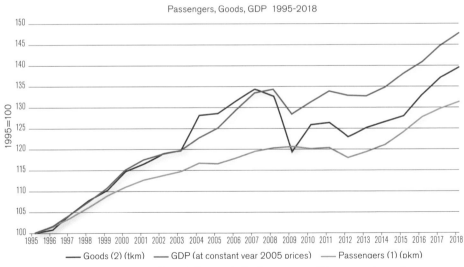

Passengers, Goods, GDP 1995-2018

Correlating Trends Between Growth Economic Performance and Transport demand

Next to economic growth we also see an urbanisation trend worldwide. More and more people will be living in cities. More people mean more demand for mobility, but also more demand for goods transport, more building materials, more goods to be transported to the shops, and so on. In cities we see the trend towards smaller shops (for example, think of the GB Express and City Delhaize) with less stock that have to be constantly supplied. And then there is the whole trend of e-commerce. Here, the delivery of goods is completely fragmented, and it is more than just online shopping. Nowadays you can buy everything on the Internet; food from a restaurant a few kilometres away, marijuana from the nearest dealer and even little kittens via UberKITTENS that you can play with for half an hour before they are taken to the next customer[27]. Online purchases are on the rise and is expected to experience further growth in the future. Moreover, COVID-19 and the related lockdowns enhanced the growth in online sales around the globe. In Europe, e-commerce volumes increased by 230% during the Christmas shopping peak in 2020 compared to the same period in 2019[28]. The lockdowns had three effects: an increase in online sales, an increase in people starting to buy online and an increased preference for home deliveries instead of out-of-home delivery locations like collection points and lockers. Most of the parcels go to the cities and research indicates that out-of-home delivery locations are to be preferred from a sustainability perspective in these urban environments[29]. Next to that, free and fast delivery offers are still part of commercial strategy of e-retailers. These two leads to inefficiencies and social malpractices in the supply chain. More on the impact of this trend is discussed in the awareness section.

So, for both passenger and goods transport there is a growth in demand and this growth is mainly met by road transport. In the EU, on average, 83% of journeys are made by car, a figure that has changed only slightly over the years[30]. Despite the growing popularity of rail travel, the modal share of trains hardly increased from 7.5% in 2000 to 8% in 2019, and then fell sharply during the Corona-year 2020 to 5.4%. During the same period the modal share of buses and coaches even decreased from 10.5% to 9.5% in 2019 and then dropped to 7.4% in 2020.

In the case of freight transport on the European land (EU27), 73% of the goods are transported by road[31]. Followed by 16% by rail, 6% via inland waterways and finally around 4% of the goods are shipped via pipelines. The shares have hardly changed over the past decades. There are large differences between countries; rail has larger shares in the Baltic states (up to 70%), Scandinavia (Sweden 30%, Finland 26%) and in Central (Switzerland 35%, Austria 27%) and Eastern Europe (Poland 22%, Hungary 25% and Romania 26%). Inland waterways are heavily used in the Netherlands (40%), Bulgaria (30%) and Romania (27%). The differences as well as the relative inertia of the modal split is explained by the presence of infrastructure (rail, inland waterways, pipeline), the huge costs related to infrastructure development, the local altitudes and policies, the economic structure of countries and the focus of the nation[32]. An additional factor in the transport of goods due to globalisation. Globalisation, partly made possible by cheap transport costs (container transport), means that goods travel enormous distances between production and consumption. Much of our production has shifted to low-wage countries. So not only is there more demand, the distance over which goods must be transported has also become much greater. The goods that are then brought to our seaports from Asia, mainly by container ships, are all too often taken by truck to the further hinterland. Also, just-in-time deliveries, such as in the automotive sector where smaller and more frequent deliveries are encouraged, mean that freight vehicles are often not full.

On the other hand, the energy consumed per vehicle kilometre has fallen slightly (by 1% for heavy goods transport and 3% for passenger transport). This decrease is mainly due to improved technology in the current fleet; the newer cars consume and pollute less. Also, there is a clear shift from diesel cars to petrol cars when purchasing new cars. This is mainly because people have become aware of the health effects of diesel cars (more fine dust and nitrogen oxides) and, following Dieselgate, they are convinced that something really has to be done about it. This shift does not help the climate at all, as they are both fossil fuels and petrol consumes more than diesel and therefore produces even more CO_2 emissions and on average there is 1 car per two persons in Europe.

Already 24.5% of new car sales are alternatively powered vehicles (battery electric, plug-in hybrid, hybrid and alternative fuels[33]). In general, the battery electric vehicles account for 9,9% of the new car purchases, whereas the plug-in hybrids for 8,7% and the hybrid ones for 22,6%. Petrol still accounts for 38,5% and diesel cars only for 17,3%. There are big differences between the different European countries, with Norway as an exemplary case, where the market share of battery electric vehicles already passed 70%, Iceland with a 33% market share and then followed by the Netherlands with 20%[34]. However, only 4.6% of all passenger cars on the road today are alternatively powered. It takes long to make the transition as on average a car is used 11,5 years[35]. There are already 243 million cars are on the road in the EU today and between 2015 and 2020 there was a 10% growth[36]. During the last COVID years, car sales dropped but only because of consumer risk perception and the delays in the production of cars due to the shortage of semi-conductors. Just like the trend we see in car sales, we also see continued growth in the number of SUVs being sold lately, even up to 50% of all the sales[37]! SUVs are quite problematic in several ways: first in terms of the climate crisis, it is clear that these cars need more material, and are thus less energy efficient. Due to their volume, they also take more space in the city and in terms of road safety, a collision with an SUV is much more damaging.

Not just a climate story: other effects of transport

When we talk about transport, we often talk about the other effects of transport, the ones that are more visible: the traffic jams, poor air quality, poor quality of space and accidents. And of course, that is also what this book is about. Because when we look at how we can move towards a more sustainable mobility for the climate, these aspects are also affected.

As far as traffic jams are concerned, you have probably already noticed it: they are getting longer and longer, and in certain places even structurally and at any time of the day. In 2017, Londoners were reported to have spent an average of 74 hours in traffic, one more hour than in 2016. Other major European cities like Paris lost 69 hours to traffic and Geneva lost 52 hours[38]. Additionally, in less than ten years (between 2010 and 2018), the number of lost hours in Flemish cities doubled[39]. We are now like boiling frogs, letting things get worse year after year and taking only a few jumps out of the pot to start moving in a different way. Aside from the time wasted, traffic jams also have an economic cost. As of 2011, 1 to 2% of the gross domestic product or about 4.2 to 8.4 billion euros is spent in traffic jams in Europe[40].

Similarly, when it comes to air quality, awareness has come late. Understandably, talking about particulate matter and nitrogen oxide is relatively abstract. Even though we see maps showing how bad things are in the whole of Europe, it remains unclear what the effect is on our health. In Belgium, a documentary reporting on how much particulate matter was found in the urine of children in Brussels and even in the placenta of women changed the awareness about the subject considerably[41]. Thereafter, the 350.000 premature deaths in Europe due to poor air quality started to speak[42]. Just compare that with the number of corona deaths, which was one million in the first year. So, more than one third but without any strong follow up in terms of policy measures!

Not only has it become clear that everyone will be affected by poor air quality in one way or the other, but it has also become clear that there is a need to involve the citizens. The involvement of so many people provide a large support base to do something about the issue. For example, citizen movements, like Cittadini per l'Aria (Citizens for Clean Air) in Italy and Filtre Café Filtre in Belgium, have contributed to air monitoring projects and consequently created a lot of awareness.

The World Health Organisation has indicated that air quality is a major problem for our public health, and in fact the biggest threat to health worldwide[43]. Air pollution cost the world a whooping sum of 8.1 trillion dollars in just 2019[44]; and for Europe, an approximate cost of 430 billion euros was attributed to industrial air pollution[45]. The biggest cause for this is still the large dependency on fossil fuels (same goes for the air quality problem!). For instance, the Flemish part of Belgium is still 72% dependent on fossil fuels for its gross inland energy consumption[46]. In addition to building heating and electricity generation, the use of these fossil fuels, especially from transport, results in emissions. Air pollution includes many pollutants, and the source varies greatly for each pollutant. Nitrogen oxides (NOx), sulphur dioxide (SO2) and particulate matter (PM) are considered the main air pollutants from transport. 55% of nitrogen oxides come from the transport sector and mainly from diesel vehicles. If we take the standards of the World Health Organisation, 97% of the European population is exposed to excessive concentrations of fine particulate matter. And that means that many people fall ill more quickly, suffer from respiratory problems, develop asthma, and so on.

Concerning the issue of traffic accidents, every victim is one too many. Sadly, a large percentage of traffic accidents victims are often walkers and cyclists even though they are less likely to cause these accidents[47]. In Europe, the number of victims fluctuates from country to country but thankfully, there has been a decrease of victims from 51 per 1million inhabitants to 44 per 1million inhabitants between 2019 and 2021[48]. Undoubtedly, reducing the number of victims is an example of the importance of policy measures. For example, the compulsory wearing of seatbelts and the improvement of the procedure for obtaining a driving licence have been crucial in reducing the number of annual traffic casualties. The realisation that drinking and driving do not go together also seems to be slowly creeping in. Like the climate crisis, concrete policy measures have been necessary to reduce the number of traffic victims, together with behavioural changes and a bit of technology.

In discussions, the local emissions of vehicles (nitrogen oxides and particulate matter) and the global emissions of vehicles (CO_2 emissions and other greenhouse gases) are sometimes mixed up. Local emissions mainly have an impact on our health, whereas CO_2 emissions do not, but they do contribute to climate change. Low emission zones in cities are often installed to tackle air quality in the city. They therefore target the most polluting cars in the fleet in terms of air quality. These are typically the older diesel cars. However, diesel cars consume less per km than petrol cars. And CO_2 emissions are directly related to consumption. Thus, you could therefore say that a diesel car is better for the climate than a petrol car. But then again, the fact is that both petrol and diesel are fossil fuels and are therefore both bad for the climate.

Thus, to say that all measures for the climate are in the same direction for a better and safer living environment may be a bit short-sighted. Nevertheless, most measures do go in the same direction. For example, think of the implementation of school streets or residential areas to increase traffic safety, air quality, better quality of space and subsequently, the quality of life. This, too, will lead to less car use and thus less CO_2 emissions.

Expectations

The demand for transport is expected to keep increasing. At the European level, the following figure is put forward:

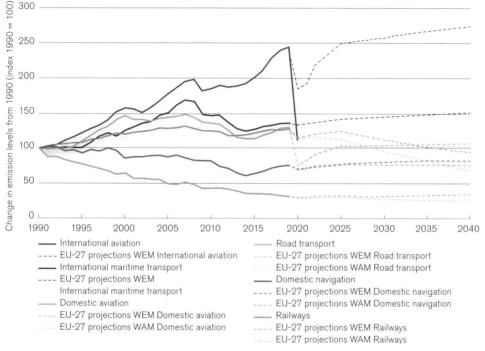

Greenhouse gas emissions from transport in the EU by mode of transport and different scenarios[49]

As you can see the railways and domestic navigation are going down, whereas road transport, maritime and aviation transport are still increasing considerably and are expected to do so even after the hiccup during COVID.

For the aviation sector you have to triple count, so it is estimated to be responsible for 5-8 percent of global climate impact. If unmitigated, aviation emissions are expected to at least double by 2050 and thus consume up to one quarter of the global carbon budget under a 1.5-degree scenario.

FACTOR 8

The challenge: a Factor of 8

The above overview shows that the current evolution, the so-called *business as usual*, is not exactly going in the right direction. Yet that direction is clearly set.

Europe has set its 2030 and 2050 climate goals and to achieve them, the transport sector has to cut its emission by 90%. This will be fully in line with the objective of keeping the temperature increase below 2°C and preferably below 1.5°C[50].

If we then make the transport target namely 10% by 2050, and we compare this with the *business-as-usual scenario* (without taking technological evolution into account), we see that a reduction by a factor of 8 is required! A factor of 8!

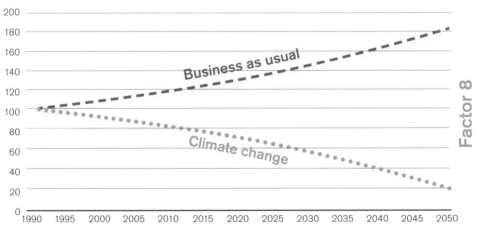

Own layout based on EU Reference Scenario[51]

So, the challenge is enormous. In this book, I indicate how we can achieve it. Clearly, we will no longer be able to get by with a few shifts in the margin; or thinking that technology will solve it. A transition to a different mobility system is needed.

So, we need to go for a transformation of our system; find another way to move around and organise logistics differently. And for that, we can use the following scheme with 8As that we will elaborate on in the next chapters. On the right-hand side, it says which actions we can take. The first one is "Awareness", so that people can become aware of the impact of their travel behaviour. The second action is "Avoid"; by asking important questions like: are all trips necessary and how can we shorten trips in the core of the system? Third, we should "Act and shift" to more environmentally friendly modes of transport such as walking, cycling and public transport, and a shift from car ownership to mobility as a service. Fourth, we should Anticipate new technologies. If we have to drive a car, let it be the most environmentally friendly car.

On the left-hand side, I will elaborate on the conditions under which the transformation to a sustainable mobility system focused on Awareness, Avoiding, Acting and shifting and Anticipation can be shaped. Here, we need to set up the right policy framework so that we can accelerate the transformation. But this also means that we will have to involve all actors, make sure there is support, make sure everyone is involved in the transformation and nobody is left behind. Besides, every transformation starts with an individual transformation. How can we ourselves alter our transport behaviour? What plays a part in that? And finally, how do we get into the right mood for a transformation? It seems to me that falling in love is the best remedy!

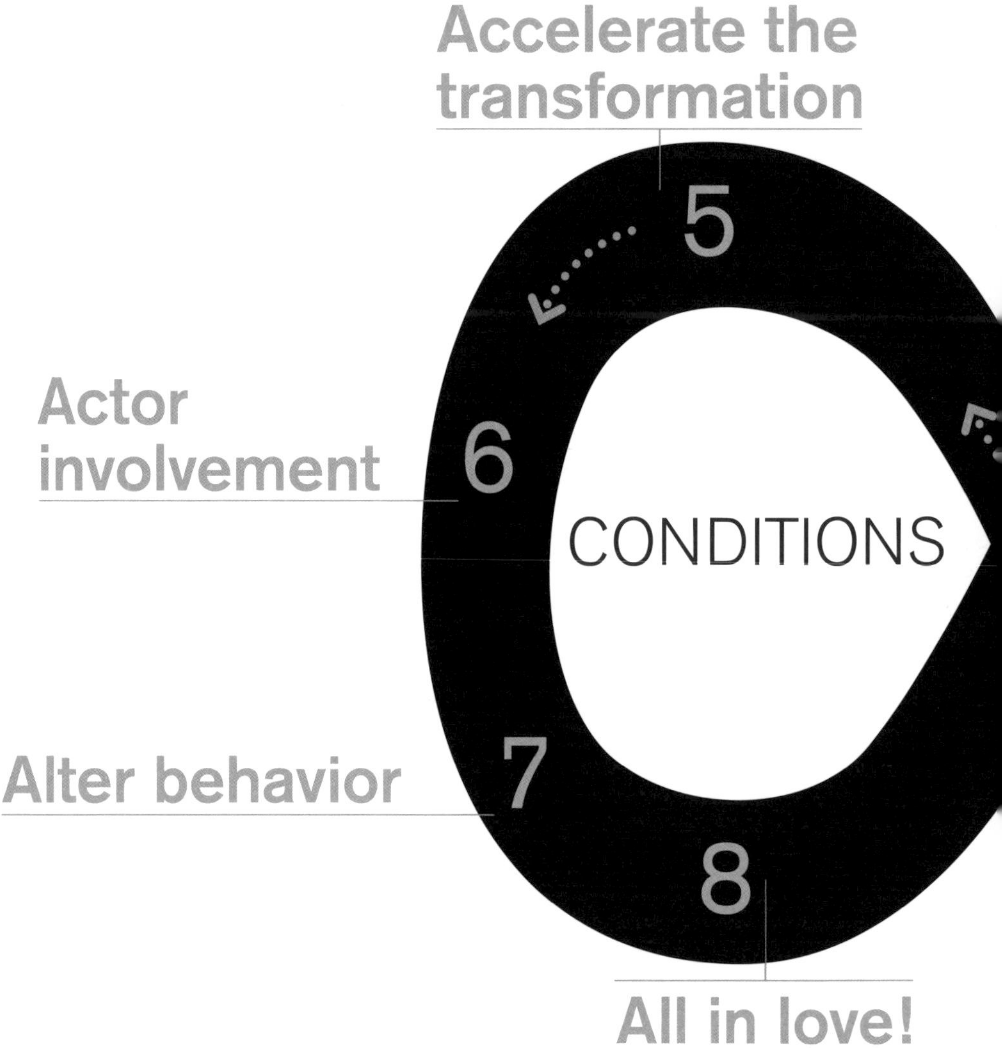

Accelerate the transformation

5

Actor involvement

6

CONDITIONS

Alter behavior

7

8

All in love!

Awareness

1

2 Avoidance

TO DO

4 3 Act and shift

Anticipation of
new technologies

What can

we do about it?

AWARENESS

How can we create more awareness?

The first A stands for Awareness: mentioning and knowing what the impact is. Becoming aware of something is indeed a necessary first step towards change, and on the climate issue, awareness is growing. According to the 2021 Eurobarometer survey by the European Commission, 93% of European citizens now see climate change to be the single most serious problem facing the world. This is a huge rise from the 2019 survey, which saw only 23% of European citizens acknowledging this. This is not just a realisation for Europe but globally as the "Peoples' Climate Vote" of the UNDP reports that 64% of the respondents (which covered half of the world's population) saw climate change as a global emergency. Lots of information is being made available and the actions of climate youngsters ('youth for climate', 'students for climate'), companies ('sign for my future') and scientists ('scientists for climate') further sharpen the awareness of the urgency of the climate problem and the necessity of sustainability.

Regarding our climate, there is no being "for" or "against" it, so any form polarisation in the social debate context must be avoided at all costs. Climate change and the influence of greenhouse gases on it is a scientifically established and substantiated objective fact. It has no political colour! It is therefore very important to report the facts objectively, about the impact of our travel behaviour but also about the alternative options that are more sustainable. Thankfully, different impact analysis methods exist for this.

CO_2 emissions per means of transport

A very simple way to weigh alternatives is to indicate the amount of CO_2 emissions per transport mode. In the following figure, this is indicated per passenger kilometre, that is, how many grams of CO_2 are emitted by a person for one kilometre of travel. Walking and cycling are of course the best, followed by public transport. And as clearly shown, air travel has large emissions per passenger, so even if you are in a plane with several people, it is still much worse than going by train. There is also a big difference depending on how far you fly. Planes emit a lot on take-off and landing, so, proportionally you have more emissions per kilometre for short flights than for longer flights, but it's a little more complicated than that. When travelling by plane, the impact on climate change is not only through greenhouse gases but also through the formation of condensation trails (radiative formation). Therefore, the effect of the insulation blanket is even greater, and for the aviation sector you have to count on an impact that is 2 to 3 times higher than just the CO_2 emissions.

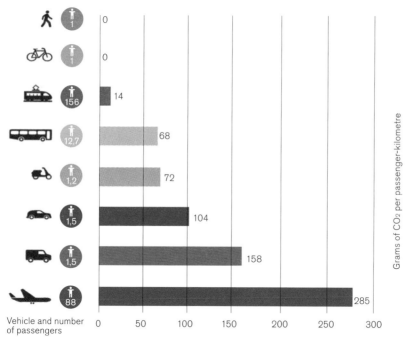

Carbon emission from passenger transport[52]

It may be possible to make it more concrete by indicating in what way, for example, the plane trips could be compensated. For instance, to compensate for the CO_2 emissions of a flight from Brussels to Auckland in New Zealand, you would have to eat no meat for 6 to 8 years or install 10 solar panels on a roof for about 2 years. However, this will not yet have removed the emitted CO_2 from the atmosphere. You will simply have

FACTOR 8

saved those emissions on your own CO_2 consumption.

How much CO_2 can we emit per person? Well, if we want to meet the 2°C target, every person on the planet should only emit an average of 2 to 2.3 tonnes of CO_2 per year. Unfortunately, this is not the case. Although in the entire African continent, the average per person is 0.99 tonnes, for high CO_2 emitting countries like Australia and United States it is 15.37 tonnes and 14.24 tonnes, respectively; and for the whole of Europe, it is 6.61 tonnes. Let's do a little comparison as there is an imbalance even within Europe. While countries like Moldova, Malta, Portugal have a per capita of 1.28 tonnes, 3.61 and 3.96 respectively, countries like Luxembourg and Czech Republic have a per capita of 13.06 and 8.21[53]. These figures are significantly greater than what is needed to achieve the 2°C target. However, it is interesting to note that some countries with high CO_2 emission per capita (emission per person) have a small population size and this factors into the calculation. Nevertheless, we can see that the figures are alarming.

In Flanders, the footprint is about 12 CO_2-eq/t per inhabitant. If you also include the production of consumer goods abroad, then we are talking about 20 tonnes of CO_2-eq per inhabitant per year, which is almost 8.6 times as much as the impact we are allowed to have. A factor of 8, so to speak!

If we look a bit deeper into this, we can see that passenger transport takes up about 2.9 tonnes, food 2.8 tonnes and accommodation 5.8 tonnes, and the rest comes from various sources such as clothing[54]. A roundtrip to Brussels-New Zealand already emits 6,6 tonnes[55]. And as earlier mentioned, for flight you have to multiply this with a factor two to three to get the impact on the climate change. Such figures put things into perspective.

In the following figure, you can see how aviation compares to other actions so you can choose to lower your CO_2 footprint accordingly.

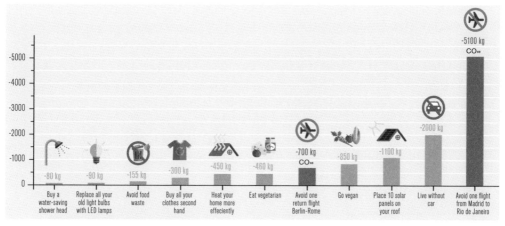

Living sustainably by avoiding flying
Greenhouse gas emission savings of different sustainable lifestyle changes, in kg CO_2-equivalents/year, 2020

And the sad fact is that the wealthiest 1 percent of people emit double the combined climate pollution of the poorest 50 percent. This, of course, is not a reason to not think about your own footprint, but it is apparent that more awareness has to be brought to these people. Flights are certainly an aspect in their lifestyle that adds to this, but also the way they invest their money and the energy use in their houses and cars.

The ABC travel policy of the Free University of Brussels (Vrije Universiteit Brussel -VUB)

Academics are keen travellers. After all, their research work has to be tested against the work of other researchers. Conferences, study trips and teaching therefore involve a great deal of air travel and, consequently, also CO_2 emissions.

For the VUB, this amounted to some 34,869 tonnes of CO_2-equivalents per year and the number is on the rise. To raise awareness about the impact of flying, the 'ABC policy' was introduced in the academic year 2019-2020. A stands for 'Avoid': do we really need to make the trip? Can't it be done via Skype? B stands for 'Book an alternative'. The train, for example. For all train journeys lasting less than 6 hours, it is recom-

FACTOR

1 AWARENESS

mended to take the train instead of the airplane. And C stands for 'Compensate': CO_2 emissions are compensated via so-called green certificates (gold standard) e.g., Greentripper from CO_2 logic. You pay about 12 euro per ton of CO_2. A trip to Marrakech and back costs you an extra 13 euro. The money is used to set up projects in Africa to reduce CO_2 emissions. For example, forests are planted, cookers are bought that are more energy efficient than wood fires and thus prevent deforestation. The amount is quite low compared to the value of a tonne of CO_2 used in economic studies today, but it is certainly better than doing nothing.

Despite such information, it is still often unclear what the impact of our actions is. For example, think about e-commerce. Is it better to shop online, or to go to a physical shop? The answer to this question is not so straightforward, and if I put the question to a group of people, most of them will say that it is better to go to the shop. In theory, however, having goods delivered to the home could be more sustainable than having everyone go to the shop by car. Such a 'milk run' with a well-stocked electric delivery van is less damaging to the environment than all the people in the neighbourhood going to the shop individually[56]. In practice, however, online purchases are often extra, and come on top of the purchases (and therefore trips) we make in the shop. Moreover, the web shops have spoiled us with free deliveries (and returns), which often take place the very next day or even within two hours. In addition to the quick delivery services that logistic services providers offer, the fact that people are often not at home to receive their package (10 to 20% missed deliveries on average) presents a sub-optimal situation. On top of that, consumers still go to the shop to research the existing assortment and test products, only to order the product of their choice online after all (the 'omni-channel behaviour'), which increases their CO_2 footprint again[57].

The extent to which e-commerce is less or more sustainable than the traditional retail model therefore depends on choices made by consumers, web shops, and logistics service providers[58]. The three actors can adjust their behaviour to make e-commerce deliveries more sustainable, by avoiding missed deliveries, stimulating manned or unmanned pick-up points near consumers, cancelling unnecessarily fast deliveries so that they can be bundled better, delivering with zero-emission vehicles (such as cargo bikes or electric delivery vans) and improving cooperation and consolidation. Governments can facilitate sustainable logistics solutions by applying adequate supporting policies; like vehicle restrictions in city centres to enhance the use of cargo bikes or by providing subsidies and/or charging infrastructure for electric vehicles.

This example shows that awareness based on hard objective figures requires a great deal of study. For example, it is often difficult to choose in the supermarket between vegetables from further afield and those grown here in greenhouses. It is not always clear which is the most sustainable choice. Nevertheless, a life cycle analysis provides the necessary input to support sustainable decision-making by all actors.

Life Cycle Assessment

A life cycle analysis considers all aspects from the extraction of raw materials to the use of the product and what is done with it after its end of life. In the case of a vehicle, this includes the production of the vehicle and possible recycling afterwards, as well as emissions during the extraction and distribution of the fuel[59].

A distinction is thus made between the emissions generated by the energy production, the emissions during the use of the car and the production of the car.

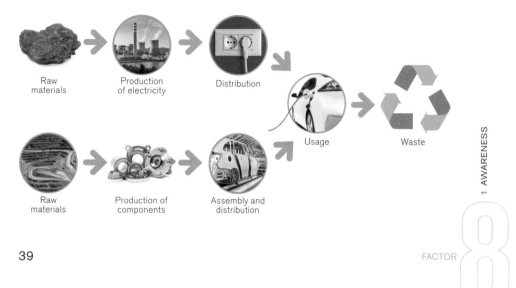

Raw materials — Production of electricity — Distribution — Usage — Waste

Raw materials — Production of components — Assembly and distribution

FACTOR 8

In this way, a more correct estimation of the impact of a product on global warming and other possible impacts is made. The figure below shows a life cycle analysis of different vehicle technologies.

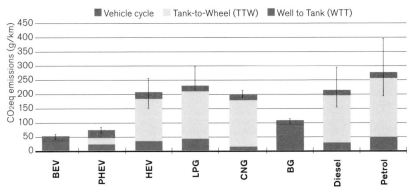

Climate impact potentials of different vehicle technologies (in CO 2 equivalents: TTW represents Tank-to-Wheel and WWT represents Well to Tank)[60].

Over their lifetime, battery electric vehicles (BEV in the figure) emit three to four times less CO_2 than conventional fossil fuel vehicles, and this takes into account how electricity is generated. So, an electric car is absolutely the best choice in terms of CO_2 emissions over its entire life cycle, including battery production.

The figure also includes plug-in hybrid cars (PHEV), cars that also run on diesel or petrol combined with electricity and that can be recharged from a wall socket. They score relatively well in the figure. However, a lot depends on the extent to which the user plugs the car into a wall socket. Experience shows that urgency and discipline are sometimes lost with this type of car, because even when there is no electricity, the car continues to run on fossil fuel.

HEV stands for hybrid cars (Prius type). Compared to pure electric and plug-in hybrids, these cars have quite a lot of emissions when driven.

LPG *(liquefied petroleum gas)* is a mixture of propane and butane. It is produced during the extraction of natural gas and petroleum and is therefore a fossil fuel. Hence, the CO_2 impact is considerable.

CNG is compressed natural gas. Natural gas is also a fossil fuel, so this type of vehicle also has high *tank-to-wheel* emissions, that is, the emissions released while driving. This calculation also accounts for the methane leakage during the extraction of natural gas.

Electric cars therefore come out best from the analysis. If they were to run entirely on green power, we would see 15 times less CO_2 emissions compared to diesel cars. Today, the electricity mix still includes nuclear energy and fossil gas. Only a small part is solar and wind energy. The more that is shifted to renewable energy, the better the comparison of an electric car with fuel cars becomes. You would have to produce all the electricity from petroleum or coal to have as much CO_2 emissions as a diesel car!

The above analysis focused on CO_2 emissions. For local emissions too, electric cars are much better. There are no emissions at all when driving, and looking again at the whole lifecycle, they emit four times less fine particulates and 20 times less nitrogen oxides (NOx).

It is important to note, however, that improvements are still needed in the extraction of raw materials in South America, Africa or China and that recycling of materials can further reduce the environmental impact. Cobalt, the raw material for the development of the batteries, is an important point of attention here. Cobalt is not always mined under humane conditions and its price is very volatile. Research is pushing hard to develop batteries with less or no cobalt. The latest batteries already use much less cobalt for the same efficiency, and the cobalt can already be recycled well since recycling the batteries and reusing the precious materials is extremely important. As with other electronic devices, proper recycling mechanisms will need to be installed.

Finally, biofuels are often used as a fuel for cars. The preposition bio here refers to the fact that the energy that is generated comes from organic material on the surface of the earth, such as wood and plant remains, but also from dried faeces, vegetable oil and animal fat. It can therefore take different forms. The advantage over fossil fuels is that burning biomass releases as much CO_2 into the air as it had absorbed during its lifetime.

With fossil fuels, all that is released after so many millions of years is extra CO_2. With plants and trees, you could say that we just let the CO_2 back into the air that the plants had first extracted themselves. The comparison is not entirely correct, because if you start to grow plants for this specific purpose, you will also need fossil fuel powered agricultural machines and you will need to transport the biofuel as well, which again means extra CO_2 emission.

Also, if biomass needs to be mass-produced to make biofuels, then the land cannot be used for other agricultural crops, which influences food supply and food prices or leads to deforestation.

We are all familiar with liquid biofuels from the pump. The E85 label on petrol, for example, indicates that ethanol has been added to the petrol. Such bioethanol comes mainly from grain, maize and beet. In the case of diesel, it is mainly rapeseed oil, soya oil and palm oil that are mixed in. Currently, the percentage of blending varies across countries but usually ranges from 5% to 10% (E5 and E10 fuels) and the European government would still like the percentage increased because this could reduce CO_2 emissions from the transport sector[61]. However, there are significant drawbacks to biofuels. Half of what we import comes from South America and Asia. For instance, Belgium produces only 3% of its biofuels. Thus, if Belgium were to increase its blending percentage, we would need a land area equivalent to the provinces of Antwerp and Flemish Brabant to do so. This already gives an idea of the enormous impact this has on the food supply both in the South and in Europe, if done on a large scale. Moreover, it is not true that biofuels would be climate-neutral anyway. Biodiesels emit on average twice as much CO_2 as regular diesel.

Biogas (BG) is a gas that results from the fermentation (an anaerobic process) of organic material such as manure, sewage sludge, activated sludge, organic waste, grass, maize, glycerine, etc. It consists largely of methane and is often referred to as biomethane. There are cars already running on biogas, especially in Sweden. The emissions during driving are compensated by the absorption of CO_2 during production. However, during the extraction and production of biogas there can be a lot of emissions due to methane leaks. If you consider that methane is 25 times more harmful to climate change, then you realise leaks are a real problem. Moreover, excessive demand for agricultural crops can affect land use and food prices.

The 'advanced' biofuels would not have these disadvantages. They are not made from food crops. Some of them are made from algae or seaweed, which puts less pressure on land use. For biofuels made from waste, the comparisons are more interesting, but here too one has to watch out for leaks.

One should therefore be careful with the excessive use of biofuels for the transport sector in order to meet the climate objectives. It certainly has a role to play, but not a leading one, because then the negative effects start to weigh out the positive ones[62].

And what about messages stating that the new diesel cars are just as environmentally friendly as electric cars? Yes, some new diesel cars with Euronorm 6d score well in the test in which the emissions were measured while driving. For instance, a Mercedes with a particle filter and selective catalytic converter even scored 10 out of 10. But the environmental friendliness referred to here is only in terms of local emissions, nitrogen oxides and fine dust. This solves nothing in terms of global emissions, CO_2. Diesel and petrol cars will always emit more CO_2 since the technology is based on fossil fuels. Of course, it is to be welcomed that these new cars have lower local emissions, but that will not solve the climate crisis.

We have not yet discussed hydrogen cars. There is also talk of hydrogen as a 'zero emission' solution, but hydrogen must also be produced. Hydrogen produced from natural gas (which is what we are doing today) has no advantage in terms of greenhouse gases as natural gas is a fossil fuel. So, you have to generate the hydrogen gas with renewable energy sources, via wind, solar or hydro power in order to have a positive story. The problem is that you need three times more wind turbines to run a hydrogen car than to use that electricity for a battery-electric car because a lot of energy is lost in all the operations. So, hydrogen has potential,

FACTOR 8

but not so much for passenger cars. Possibly for maritime transport, as battery-electric sailing is not interesting there. We will come back to this in the chapter on "Alter behaviour".

Putting a monetary amount on emissions: external costs of transport

Another way to create awareness, especially among governments, is to place a monetary value on the cost of transport to society. These are called external costs because they are not included in the price we pay. As we saw earlier, apart from the impact on the climate, there is also a significant impact on health due to air pollutants, time loss from traffic jams, accidents, noise pollution, and so on. The European Commission estimates the total external costs of transport in the EU at around 1,000 billion euros a year, or, as almost 7% of the EU's GDP28[63]. That is huge!

The way in which prices are put on all these different effects has been studied for a long time and there is now mostly scientific consensus on the methods used[64]. For example, by estimating the loss of productivity due to pollution-related illnesses or the medical costs related to health problems caused by fine dust or by having people value the time lost in traffic jams and see what they are prepared to pay for it.

For Brussels, we calculated that there is a daily external cost of €51,692 and that this is solely attributed to the effects of fine dust and nitrogen oxides, that is, local emissions, caused by goods transport in our capital city[65]. The concept of external costs is somewhat difficult to grasp, because who pays for it? We all do, society as a whole. Because, every day, health problems are created by fine dust and, as a result, there will be more medical costs, loss of production, and so on, which we have to pay through our social security system. So, we'd better remove the cause directly and invest the money in avoiding the costs, right?

However, with respect to climate, there is still a lot of debate about setting an accurate price for a tonne of CO_2. After all, it is very difficult to set an accurate price for all the damage caused by climate change. How do you put a price on the loss of biodiversity, the acidification of the oceans, rising sea levels, extreme weather conditions, drought, hunger, the spread of tropical diseases, political instability and other effects that we have not even thought of yet?

You could do that from different perspectives. What is all this damage going to cost us, or how much money are we prepared to invest to avoid certain effects? More and more people are looking at it from the perspective of: what should the price be to meet the 1.5°C targets?

Based on the latter question, a commission led by economists Stiglitz and Stern calculated that a price of €35 to €65 per tonne of CO_2 in 2020 and of €40 to €80 per tonne of CO_2 in 2030 is needed to achieve the Paris objectives[66]. The Intergovernmental Panel on Climate Change (IPCC) suggests a price between €18 and €90 per tonne of CO_2. Other studies find these amounts too low. The latest update of the Handbook on External Costs of Transport[67] gives a central value of €100/tonne CO_2 equivalent. This is the amount often used in studies, but in more recent studies even higher amounts are used[68].

Why is it important to put such amounts on CO_2 emission and to do it correctly? The fact that the externalities are not paid for means that too much of it is consumed. That is a simple economic principle of supply and demand. And as a society, you lose prosperity as a result. But, by integrating it into the price (internalising it, as it were), you counteract this overconsumption. So, if all processes that emit CO_2 are taxed with a CO_2 tax, you will automatically consume less of it. Many countries are already introducing CO_2 pricing[69]. Many companies already use the cost of CO_2 in their calculations to factor in the possible introduction of CO_2 taxes at a later date. However, 80% of the emissions are not yet taxed.

The current price of a tonne of CO_2 in the European Emissions Trading Scheme fluctuates between €20 and €25. This system has been widely criticised because it is a system of supply and demand, and the price goes down when there is a lot of supply. It therefore provides insufficient incentive to invest in climate-neutral technology.

Another way to evaluate the costs related to CO_2 is the carbon offsetting mechanism, according to the principle of CO_2-neutrality. The mechanism of CO_2 compensation states that an organisation or a person compensates (part of) the greenhouse gas emissions it produces by paying for a CO_2-equivalent reduction in another region of the world. For example, a company might invest indirectly in wind farms and thus compensate for a CO_2 saving equivalent to the emissions from coal-fired steel production. Or you can offset your plane flight to New Zealand by co-financing energy-efficient cookers in Uganda; these appliances reduce the amount of charcoal used by households in the preparation of their daily meals, thus contributing to a reduction in CO_2 emissions. Using this approach, climate neutrality is seen as reducing emissions elsewhere on earth but lower amounts per tonne of CO_2 are often used for these calculations, such as €12/tonne. If it were really to be seen as a CO_2 tax, the amounts would have to be higher, as indicated above. I see it as a pragmatic interim solution whereby conscientious people can still contribute to projects that have a positive effect on CO_2 emissions but are not yet charged the full consequence or price of the CO_2. Nevertheless, it is important that the projects are certified and thus checked for authenticity by a supervisory body such as Gold Standard, which was set up by the World Wildlife Fund, among others[70].

In general, we can say that transport is too cheap because the external effects such as air pollution, accidents and impact on the climate are not included in the price and therefore we use it too much. Moreover, the taxes that do exist are not sufficiently directed towards those means of transport that are the most polluting, especially road transport[71] and certainly not aviation. We will come back to this in the Acceleration chapter.

Let us first see if we can avoid certain impacts.

FACTOR 8

1 AWARENESS

AVOIDANCE

Once we become aware of a problem, we can do something about it. And the best first step towards more sustainable mobility is avoiding unnecessary kilometres. To achieve this, we need to examine spatial planning. After all, we move around because we have an activity, such as work, study, shopping, and so on, that does not take place where we live. The same principle applies to the transport of goods: the closer our point of consumption to the production site, the fewer kilometres we have to drive. Secondly, we can avoid travelling by working at home or organising meetings via video conferencing. Thirdly, we can also avoid kilometres and make better use of journeys by sharing them.

Spatial planning: Reducing the need to travel

The distance between the locations of our activities such as living (our homes), working (offices, factories), studying (schools), shopping (shops and services), spending our free time (sports fields, restaurants, bars, cinemas) are one of the key determinants of how far we need to travel. If the locations of these activities are very scattered, more trips are needed to live our lives. Spatial density defines the distribution of these activities and the denser our urban areas are built, the higher the chances that you can find a shop, school or a job closer to where you live, hence avoiding long trips.

In Belgium, for example, activities are extremely fragmented. If you look at the area around Brussels (Figure 2, centre) compared to what we observe in the same area around the Randstad in the Netherlands or Paris, you can see that activities are sprawling across the metropolitan area and beyond.

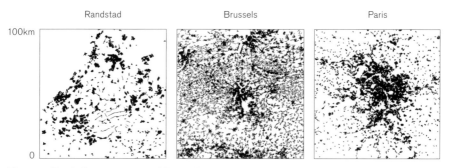

Figure 2: The built-up area around Brussels is much more fragmented than around Paris and in the Randstad[72.]

This fragmentation (also known as sprawl) has many consequences. First, it makes it very difficult to offer a high-performance public transport system. The public transport network therefore leaves much to be desired. And this also means that the car dependence of people who do not live in the core city is much greater[73]. If you look at the large densely built cities with a good public transport network, the modal share of the car can be as low as 15%[74] with a high percentage of trips made by public transport, walking and cycling. In areas with fragmentation and sprawl, like the metropolitan areas around Manchester and Birmingham

FACTOR

2 AVOIDANCE

in the UK, the car accounts for 65-70% of trips.

Secondly, because of the fragmentation, people tend to live further away from their work. In Belgium, for example, people drive 6% more kilometres to work than the Dutch and 9% more than the French. For people with a company car, that distance is even significantly greater, namely 32 km compared to 19 km[75].

These two elements together mean that the social and environmental impact of mobility in an area with low density are twice as high as in the city centre[76].

Spatial planning, that is, the long-term planning of activity locations is thus a crucial element to influence mobility and its consequences at the source. The most important fundamental layer of a sustainable mobility policy is therefore spatial planning. Here, we not only need to make our metropolitan areas, cities and even communes denser, but we also need to do this in a smarter way, namely by working with mixed functions. If living, working, shopping and relaxing can all be done in the same place, why move around? In practice, there are already fine examples of such mixed-use developments, such as at King's Cross in London, where one of the largest mixed-use development projects in Europe has been implemented, combining office, retail, service and residential spaces with excellent public transport connections.

Urban sprawl has many other effects as well. Admittedly, just like many other people, I find a house in the greenery quite a nice dream. But for society, it gives rise to too many additional costs. It means that people are more car dependent, but also that more infrastructure works are necessary, that the house needs more heating, and so on.

If the costs of infrastructure (maintenance of roads, utilities such as water, gas, electricity, sewerage and lighting), the loss of open space and ecosystem services are also calculated, then you get large differences between that house in the green countryside and an apartment in the city. All in all, by choosing a scenario in which open space is given back to nature, it was calculated, that for Flanders, 25.6 billion euros can be saved by 2050[77].

Interestingly, if we leave the countryside and go back to the city, improve the limited public space within the urban area by reducing the number of parked cars, add more greenery, build bike and walk infrastructures, we can actually have the living environment that we are searching for outside the city.

Teleworking and teleconferencing

Many commuting trips can be avoided by allowing people to work at home or at a satellite office nearby. Thanks to teleworking, a lot of external transport costs can be avoided. For a day worked at home per week, we are talking about a 20% reduction in external costs[78]. It is often said that people who work at home will still travel that day, but this does not outweigh the home-work trip. Our research shows that only a very small proportion (5.4%) of teleworkers use their car for extra journeys while working at home. In addition to the benefits of reducing external costs, working from home can contribute to the wellbeing of employees by reducing time spent in traffic jams during peak hours and provide increased productivity and financial savings for both the employer (less office space required) and the employee (reduced commuting)[79].

Working in satellite offices, that is, an office location of a company separate from its headquarters, can also have a positive effect. However, if one does not consider whether the employees can also travel to the satellite office with the same train pass, it can cause unexpected impact on travel behaviour. For instance, our research showed that people who normally commuted to work by train to the company headquarters, changed to car when commuting to the satellite office. This was not so much due to the poor accessibility of the satellite offices by public transport, as satellite offices are often very well connected, but more because the travel was not included in their public transport season ticket, and they therefore chose to go by car. Coworking spaces can also be interesting for digital nomads who have the freedom to work anywhere, but also like to do so with people around them.

2 AVOIDANCE

Many meetings or appointments can also be done via videoconferencing, using Teams or Zooms. During the corona crisis, we all learned to work with these apps, which means a lot of – often long-distance – meetings can be avoided. At the university, we encourage people to use it more. For example, evaluation commissions for PhDs and some conferences are now organised virtually. Of course, we remain people and real (eye) contact is important for networking, but some of our activities can certainly be done just as well without travel. Furthermore, lessons, doctor visits, but also hobbies such as dancing, meditating... just about everything has been tried out digitally during the corona crisis. As of now, many employees and employers voiced they want to keep a certain amount of telework and teleconferencing, even after corona. Tuesdays and Thursdays are popular days to go to work while the other are often used for teleworking. The public transport operators also notice this in their capacity use and are adapting their passes to make them more flexible.

Sharing trips: avoidance through more optimal use of capacity

Look around when you are in a traffic jam, there is often only one person in the cars next to you. And you are also probably alone in your car, even though there is much more room in the car. Car occupancy is an indicator of the number of people travelling in a car. Average occupancy rates in Europe range between 1.1 and 1.2 for commuting to work and 1.4 to 1.7 for family trips[80], while the capacity of passenger cars is typically 4 to 5 persons. We can avoid a lot of journeys just by driving together[81]. While a few years ago we had to stick our thumb in the air to ride with others, or use a company organised carpool system, thanks to new technology and smartphone apps, there are now many other possibilities.

However, carpooling for home-work trips still hovers at 3% of trips[82]. Applications like BlaBlaCar are becoming more and more popular in Europe. They make it possible to find people who have a seat free for the destination you want to go. So, the platform connects people who were going to travel anyway. The people who offer a ride do not do this as a professional activity and only the cost of the ride is shared.

Uber or Lyft work based on a different business model as these are taxi-like services where drivers and passengers are linked in real-time via a mobile application based on a specific demand for trips and supply of vehicles and drivers (it is often called ride-hailing). The impact of these services on other modes of transport is still unclear. A study in the San Francisco region shows that 31% of the trips made by ride-hailing replaced trips by public transport[83].

In itself, a concept like Uber can help reduce car ownership in the city[84]. However, we have to watch out for the long-term effects. In the US, for example, certain public transport services are replaced by Uber services. Another example from the US is the construction of parking spaces at a train station. These are no longer needed because Uber can take care of the trip to the station. In the short term, you do save money that can be used to subsidise the train passengers' Uber services, but what if Uber raises its prices substantially after a while? They will already have a monopoly and the money to build the car park will no longer be there.

Illustrative of what is to come is the evolution in New York. There too, Uber and Lyft made their appearance with fine promises about the positive effects on mobility: more shared mobility instead of solo use of cars and more efficient services than the existing taxi industry. In 2015, it was mainly about attracting the customers of the yellow taxis, but gradually it was noticed that there was more and more congestion in New York. As it turned out, people who used to take the bus and underground were also increasingly starting to use Uber services. These trips were not so much shared services, but solo and this led to many more rides in an increasingly congested city[85]. The idea that Uber is especially useful where public transport is temporarily not possible because of too little demand is nice in theory, but in practice the Uber drivers also work where there is the most demand, usually in the centre of the city and not in the inaccessible zones. Such services must therefore be closely monitored in order to ascertain whether or not they contribute to the city's sustainable mobility objectives.

FACTOR 8

Avoidance in freight transport

Avoidance in freight transport seems a lot more complicated than in passenger transport. In the end, we want to get those goods to a final destination so that we can use them, don't we?

But here, too, much can be avoided. First, by making the products smaller. For example, for soft drinks or fruit juices, it is not necessary to transport all the water. You can add that at the end. Washing products are also becoming more concentrated, so that less is needed and therefore less is transported. Some products are even completely dematerialised, such as e-books, films and music, so that they no longer need to be transported.

A second focus of attention may be packaging. Ikea, for example, has ingeniously improved the way furniture is transported. In this way, a lot more furniture can be transported in the truck and the displacement of air is avoided.

In the early years, the packaging of online orders was a real crying shame. By using boxes that were far too big, a lot of air was displaced besides that small object you had ordered. Nowadays, this is much more measured, and packaging is better suited to the content. In the UK, the company Bloom&Wild, which delivers flowers, has adapted the packaging so that it fits into most letterboxes. This not only avoids the need to make extra trips because nobody is at home, but also avoids transporting too much air.

BLOOM&WILD *the flower Journal*

Available soonest   Peonies new   Letterbox flowers   Hand-tied bouquets   Plants   Subscriptions   Occasions   Business   All

So, how does it work?

We pack our flowers by hand
They're picked in bud and wrapped in petal protectors so they travel safe.

We pop them in the post
They cleverly fit through the letterbox so they can be delivered if no one's in.

They get 5 minutes of fun
Every bouquet comes with fun arranging tips so they can style theirs like a pro!

Avoiding missed deliveries of online purchases is an important concern. 14% of people are not at home at the time a parcel is delivered[86], resulting in extra trips. In addition to the pick-up points and locker systems, experiments are now being conducted with deliveries in the suitcase of the car or even with smart access to the home (via smart lockers). This way, the parcel delivery person can come and put the goods in the locker himself. New houses will also no longer have such small post boxes, but larger boxes in which parcels can be delivered easily.

So, a lot can already be done to avoid unnecessary kilometres. However, the biggest challenge in logistics is to make better use of existing capacity. 20% of trucks drive empty and if they are not empty, the load factor is on average 56% and in an urban environment only 38%. This means that there is a lot of air being driven around. To remedy this, we must look at how this unused capacity can be better filled. At Colruyt, for instance, the lorries are fully loaded because their planning process is very focused on that. If there is any space left in the lorry, they strategically take something with them that will be needed in the coming period. They also take returns such as packaging back with them.

Cooperation between companies is sometimes necessary to find optimal combinations. For example, Procter & Gamble (the heavy-duty detergents) collaborated with Tupperware to fill the trucks better. In this way, they get the lorries optimally filled and have to drive around less. Goods flows can also be combined at the edge of the city, in a city distribution centre. Building materials, too, can be bundled together and thus be distributed more optimally across the various sites in a city. From here, goods can then be delivered to the city using electric delivery vans or cargo bikes. An increasing number of logistics service providers also want smaller places in the city, so-called micro-hubs, to deliver goods quickly from there. Experiments were also made with a mobile depot, a converted truck that only had to drive into Brussels once a day, and from that hub everything was taken to its final destination by bike. In terms of CO_2 emissions and local emissions, we saw a big improvement (24% and 58.5% respectively). But this way of working was more expensive for the logistics service provider even so, that depends on the number of parcels that have to be brought to the area; if that number is higher, it becomes more interesting again.

FACTOR 8

Bringing streams together is therefore the message. Many companies are starting to work together. L'Oréal and Proximus, for instance, are bundling their goods in Brussels so that they can be delivered by cargo bicycle. Often, this requires a neutral party or a platform to facilitate this cooperation and to analyse the goods flows, because there is still a lot of distrust in cooperating, especially when it comes to working with the 'competitor'. However, more and more people are starting to realise that this can reduce the ecological footprint and that competition mainly takes place on the shelf in the physical or online shop. Sharing information about capacity in the transport network and demand is hugely important in this regard[87]. Just like the need to carpool in passenger transport, it would be best to use the existing capacity here as well.

We could also give the goods to people who are already on the road anyway. The idea of Uber but for parcels. The 'crowd', that is, us, can then take those parcels with them. Together with the VIL (Flanders Institute for Logistics), we studied the impact of this way of working for bpost. They wanted to see what the impact would be on the delivery of parcels with Bringr. People could register on their platform to start delivering parcels. Just like with Uber, the underlying idea is interesting, but in practice, things often go differently. Half of the times, people have to make a special trip to bring a parcel, 32.5% of the deliveries happens with a large diversion and only in a small minority of the times (15%) it concerns a trip that would happen anyway. Compared to how bpost normally delivers these parcels with a delivery truck; *crowd logistics* is five times worse for the climate! Of course, there are other ways to organise crowd logistics, for example by only letting people take the parcels by public transport or bicycle; train and metro stations can be interesting places to redistribute parcels.

How else can we avoid it? Here, too, spatial planning is the basis of the flow of goods. Due to globalisation, products have often seen a large part of the world. By focusing on short chains and circular economy, you can reduce the number of vehicle kilometres at the base. In the construction sector, 70% of the materials used in a house can be recycled and 25% reused[88]. And then a city suddenly turns out to be full of materials and you don't have to transport everything from afar. A lot of material or even raw materials can come from closer to home. I always give the example of the chickens in my garden. Thanks to them, I can get rid of the leftovers from the food, and I get eggs in the morning. And I only have to walk three metres.

ACT AND SHIFT

In part one, we already saw that car use still dominates our journeys. It still remains a very easy, comfortable way of getting around. What would it take to get people to travel in a more environmentally friendly way and to put freight on more environmentally friendly modes?

■ Space for cars   ▨ Space for peds   ■ Space for bikes   ■ Dead space   ▨ Buildings
▨ and used space   ■ Peds crossing   and used space

The arrogance of space[89]

A historical perspective: a story of space

A documentary about how Leuven evolved from a city that, like many others, became saturated with cars and then pursued a policy to make the city car-free again is an eye-opener[90]. It shows how streets that used to be meeting places for neighbours and friends, and where children could play carefree in the street, have been completely taken over by parking and cars driving around with unhealthy exhaust fumes. The installation of the first shopping street, Diestsestraat, caused a lot of protest, most of all from the shopkeepers themselves. This is the evolution we have seen in many cities. From the 1960s onwards, cities were completely built to accommodate as many cars as possible. One of the reports also clearly indicates that the tram had to make way for the car. Then the focus of cities shifted from cars to sustainable mobility. A next step in urban development is to see the city as space and to give that space back to the people[91].

That is quite a difficult task, because in the meantime the car has taken such a prominent place that it is hard to change it again.

Urban planner Mikael Colville-Andersen uses colour in photographs of urban streets to show how the road surface is divided between cars, cyclists and pedestrians. He calls it the 'arrogance of space' and argues for a redistribution of urban space in favour of cyclists and pedestrians. In cities, often 50-70% of the space is occupied by the car[92]!

Why am I telling you all this? First, to show that car dominance has not always been there. And second, that we are in a negative spiral, because who wants to ride a bike in a city that is completely geared towards cars, so you have to be slightly suicidal to jump on a bike. If public transport is also poor and you get stuck in car traffic jams, you are not really motivated to jump on the bus. And do you want to walk in a city where the pavements are small, and you have to walk amidst the emissions and the crowds of cars? No! And so, we have even more people in the car. And there are people who flee the city to live in a greener environment with better air and then commute by car to work in the city. If we want to get out of this negative spiral, we have to make an effort to give that space back.

FACTOR 8

This can be done in a very clear way by installing pedestrian zones. In Brussels, one of the largest pedestrian zones in Europe was installed thanks to the citizens' movement 'Picnic the Streets'. Not only does the zone itself give way to pedestrians and cyclists, but it has also given rise to the shift we are talking about here. Now, 14.5% of the people coming to the pedestrian zone no longer come by car but rather by public transport, and there were 2.5 times more pedestrians in the pedestrian zone. Among employees working in the zone, 9% also indicated that they had shifted from driving a car to other sustainable modes[93]. Similar stories can be found in cities and other countries. The initial resistance of residents and shopkeepers quickly turns into a positive assessment of the introduced changes. And the expected traffic chaos does not materialise, because the traffic evaporates[94]. People no longer come or come with other means of transport.

So, it is about giving space back. In Barcelona, they work with the 'superblocks concept'. In such a superblock, which often consists of nine blocks of houses (see figure), priority is given to cyclists and pedestrians. Brussels, too, is already working on such a district level, so that it can become a car-free district. In fact, you restore the liveability in such a district. School streets and bicycle streets also give the street back to the active modes.[95]

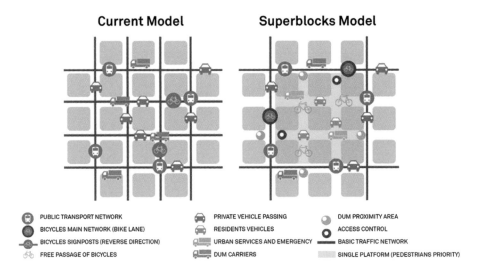

Current Model	**Superblocks Model**	

PUBLIC TRANSPORT NETWORK   PRIVATE VEHICLE PASSING   DUM PROXIMITY AREA
BICYCLES MAIN NETWORK (BIKE LANE)   RESIDENTS VEHICLES   ACCESS CONTROL
BICYCLES SIGNPOSTS (REVERSE DIRECTION)   URBAN SERVICES AND EMERGENCY   BASIC TRAFFIC NETWORK
FREE PASSAGE OF BICYCLES   DUM CARRIERS   SINGLE PLATFORM (PEDESTRIANS PRIORITY)

To enable a shift to cycling, the only option is to invest in infrastructure. By the way, that cost far less than road infrastructure, and it yields so much more. Copenhagen, the example of a city that has made the shift happen, did only three things: and that is invest in infrastructure, and again and again. Parking lanes were taken away to build free-flowing, metre-wide cycle paths. But also, all kinds of other infrastructure interventions, from tunnels to smart traffic lights, ensure that there is as little friction as possible between cars, public transport, cyclists and pedestrians. Bicycles are also allowed on trains and subways which makes the combination easier. This transformation into bicycle cities can also be seen in many other European cities.

Cities such as Copenhagen and Amsterdam, which cannot immediately be labelled as the sunniest compared to our region, still appear to get many people on their bikes thanks to stimulating measures. Once the tipping point is there, thanks to sufficient infrastructure and sufficient attention and space for active modes, cycling and walking become an automatic reflex. And as the Danes say: there is no such thing as bad weather, only bad clothes. So, investments in safe cycling infrastructure pay off. Our research into why people do not cycle in Brussels also revealed that, in addition to the infrastructure, individual and social factors also play a major role in whether or not people cycle. For instance, cyclists receive more social support from partners, colleagues, friends and/or children in their decision to cycle. They also had a higher score for self-reliance, which means that cyclists are not discouraged from cycling even when it rains during the journey to work, for instance[96]. It is, as it were, part of the culture and a way of life.

With electric bikes and speed pedelecs, cycling is now also becoming interesting for longer distances. In 2018, 503,119 new bicycles were sold, of which 50% were e-bikes[97]. With the corona crisis, the bicycle revival has really taken off. After all, it is both healthy and safe and additional cycling infrastructure was provided.

In addition, you must ensure that there is a quality supply of public transport[98]. And therein lies one of the biggest problems. Public transport requires major investments, but instead of investing in it, it has just been reduced in many countries. The Brussels region shows that things can be different, where there has been investment in public transport and cycling infrastructure. Research shows that these investments go hand in hand with reduced car use and even a decrease in car ownership.

Investing in more services and more frequent services is one thing, but comfort is also increased by providing real-time information, being able to transfer easily from one line to another (even if it is from a different operator) and, here too, it is important to give space and priority. A bus or tram in a traffic jam is no use to us. The corona crisis has made public transport not such an obvious choice. The idea of all of us crammed together in a bus or tram, after so many weeks in isolation, seems like the most chilling moment in a thriller. But to be honest, that's what it was for me before. That is precisely why further investment in capacity and in hygiene (why not install hand gels dispensers at the door) are important. What already became clear during the corona crisis was that, thanks to the reduced traffic, the travel speed of the overground public transport in cities increased considerably. As with cycling, the provision of dedicated bus lanes and dedicated tram tracks is very important in enhancing the attractiveness of public transport.

A clear parking policy can also bring about a shift. Making parking paid and/or restricting it has an influence on travel behaviour. The parking policy is also used to obtain more space in the streets and get more cars into underground car parks rather than on public roads. But the greatest influence is exerted by parking spaces in and around companies. Until recently, when a new office building or shopping complex was built, a minimum number of parking spaces was imposed. In the US, this number was so high that it determined the spatial structure. By reversing that logic and providing fewer parking spaces, companies are motivated to encourage their employees not to come by car.

The provision of Park&Rides on the outskirts of the city can also be a way of getting people in the city to switch to public transport[99]. Currently, several P&Rs are being constructed in the major cities in Europe and there is even an app that ensures easy accessibility.

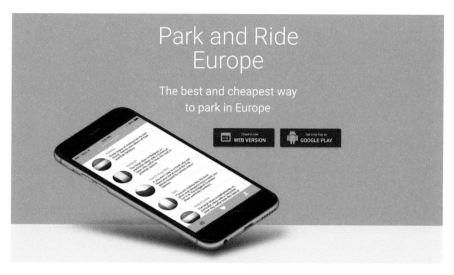

Europe's Integrated Park & Ride App[100]

However, the principle of P&R is much criticised because it often generates more car kilometres than it saves, and it also takes up space at the expense of densification around public transport hubs.

FACTOR

3 ACT AND SHIFT

Sharing is caring?

The issue of space usage becomes even more striking when you realise that on average, a car spends 96% of its time standing still[101]. Most of the time it is at home (80%) and then another 16% at other places. This leaves 3 to 4% of actual driving time.

By sharing vehicles, they can be used more optimally. One shared car can replace between 2.5 and 13 cars[102]. Moreover, it is a nice addition to public transport and can therefore make people decide to stop owning a car. People who drive a shared car are in any case more confronted with the cost of the car and it also requires some organisation... which makes them more aware and less likely to drive. There is still some debate as to whether or not the shared cars have a positive effect.

In addition to the traditional car-sharing vehicles that must be returned to a fixed location, there are now also cars that can be left anywhere in the city, such as ShareNow[103]. We see that the type of user of these new *(free floating)* systems changes compared to the more traditional way of carsharing. Based on figures for the Brussels region, we see that users of the traditional shared car systems with a fixed location are often women who do not have a car and who use the shared car for those trips that are difficult to make by public transport. In the new systems, it is young men - who often have one or even two cars in the family - who see the shared car system as an extra option in their travel profile. In itself, it is very positive that shared cars also appeal to a new market segment, but the question is whether these new users will indeed travel more sustainably. The first figures rather indicate that they use public transport even less than before[104]. It is therefore important to monitor the use of shared cars closely in order to know whether shared cars are indeed contributing to a sustainable mobility policy.

An additional advantage of the car-sharing system is that it eliminates the need to buy large cars, measured for the one day of the year when you might need to transport something large. You can easily reserve the right type of car for the trip you want to make. Besides the professional providers of car sharing systems, there are also peer-to-peer systems where people share cars locally[105].

In addition to shared cars, shared kick scooters and shared bicycles are now appearing in many cities. For shared bikes, too, there are now free-floating variants where you can leave the bike wherever you want.

These sharing systems can give just that extra push to make the first part of the trip with, for example, a scooter and then continue with public transport - according to a recent study in Brussels scooters[106]. Nevertheless, many severe accidents are being reported for the use of the scooters which is very worrying. Statistical data are not yet adequately collected, but in 2021 there were 1,000 accidents involving electric scooters across Belgium, with about 400 in the Brussels region alone[107]. What is worrying about the sharing systems is the speed with which the vehicles are replaced. Shared vehicles are replaced within two years and shared scooters within two months. In the case of shared cars, one could argue that they are always the latest models and consequently results in a greening of the fleet. With scooters and bicycles, the question is whether such rapid replacement will make the entire system more sustainable. The answer depends very much on the extent to which they help promote the shift to more environmentally friendly modes.

Mobility as a Service (MaaS)
The next step in the development of our mobility system is the arrival of Mobility as a Service (MaaS) packages. With all the sharing systems and different public transport providers, in the long run you have so many apps on your phone that you have to pay for them each time in a different way. MaaS aims to provide a solution here. Mobility as a service is the integration of the various modes of transport in one package, i.e., public transport, shared cars, shared bicycles, taxi services, charging points, and so on. With such a subscription, you no longer need to own a car, but you have the freedom to choose the means of transport that best suits the trip you are about to make. Mobility as a Service with the Whim app was started in Helsinki by MaaS Global.

FACTOR 8

They have also rolled out their system in Antwerp, and a subscription looks like this:

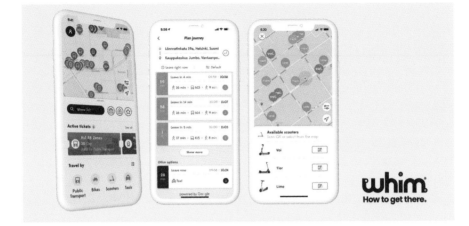

You can either pay per ride but without having to use the different payment systems of the operators; Or you can go for a subscription system. For 55 euro/month, you can have unlimited use of public transport, shared bicycles and you are entitled to a 10 euros taxi ride and credits for a shared car for the value of 49 euro/day.

There is no need to have different apps on your mobile phone, as everything is managed and paid for through one system.

MaaS users appear to use public transport more than others - 63% more users compared to 48% in general; and 95.2% of the trips are made by public transport[108]. Such a subscription clearly does provide an incentive to use more sustainable modes instead of the car. However, this is only true if they get rid of their car. If they still have a car, they tend to still use it a lot even with their MaaS possibilities[109].

Without a solid public transportation system, it seems to me that it is very difficult to introduce MaaS. MaaS global, the company who started the idea in Helsinki, is now active in Helsinki and Turku (Finland), Antwerpen (Belgium), Vienna (Austria) and the West Midlands (UK), multiple cities in Switzerland, and Greater Tokyo (Japan). Also, governments and public transport companies are now starting to make their own application to combine different transport solutions. The big question here is how to implement it optimally. The private players have an easier time integrating the technology behind the whole system in a user-friendly app. However, the government, as well as the public transport companies, have every interest in not standing on the side-lines and watching what is implemented. After all, in order to achieve an inclusive, sustainable transport system, a few preconditions need to be drawn up. For example, will the allowed vehicle kms related to car use be kept, or will we all start to drive around if we have some "rights" left over (cf. the call credit in a mobile phone subscription)? What about pricing? Will there be a distinction between premium subscriptions with very high tariffs that give you access to all options and basic subscriptions where you will only have the public transport option? And how do we reconcile subsidised public transport with a system where profit margins are levied on its use?

Towards a car-free life

A recent Eurobarometer survey showed that 59% of regular car users are interested in using a MaaS-type app[110]. This chapter showed that there are many other options than the car and that a combination of all these options gives much greater freedom than just sticking to the car. You can see it as a natural evolution of the mobility system.

I have tried to visualise this in the figure below. Before 1900, we used to travel by bike but after the car era started in the fifties and sixties, car ownership and use experienced a strong growth. The resulting traffic jams and air pollution no longer made cities liveable. Public transport is a good alternative, but it is now also being supplemented by shared and micro-mobility, which means that combinations can be made, and we can move away from a system of car ownership to a system of mobility use.

The next step in this evolution is the use of autonomous vehicles. More on that in the next chapter. But first we will have a look at the evolutions in freight transport.

Shifting in freight transport

Freight transport, too, can be shifted to more environmentally friendly modes, such as rail, inland waterways and cargo bikes.

Despite the many attempts to create a modal shift towards rail and inland waterway transport, the relationship between the three modes – road transport, inland waterways and rail – has remained the same at European level for the past 30 years. Yet, things are moving. For instance, inland navigation is paying a lot of attention to the possibility of supplying cities. Historically, cities developed next to or around a waterway. In addition to the traditional bulk transport, inland shipping has started to transport more and more containers and even pallets. There is a growing interest in this, especially for building materials. Large construction sites located near the water are now often supplied by inland waterways. Even for other types of goods, such as liquor or packages, people are looking for solutions via water. In Utrecht, for example, Bierboot, an electric cargo ship that delivers to catering establishments via the water, is in operation. In Paris, containers of Franprix, a large supermarket chain, are also transported by inland waterway. Also, in Paris, FLUDIS is transporting goods, parcels and IKEA furniture into the city by barge to various stops along the Seine. From there, the trip to the final destination is done by cargo bikes.

Cargo bikes have great potential for bringing goods into the city. Fifty per cent of what is brought into the city can be delivered by cargo bikes. They offer transport capacity between 40 and 250 kg for goods and persons. Interestingly, they are legally bicycles as long as their electric assist cuts-off at 25 km/h, have an average power of max. 250 watts and they do not exceed possible limits for dimensions and weights of bicycles in national street codes. In Brussels, these dimensions have been adapted in order to make it even more competitive. A lot of new initiatives are emerging, such as the Dutch e-commerce company Coolblue that delivers online orders by bike. Also, for personal use (owned or shared) the cargo bike has a great potential and can be seen as a solution for the last mile and for getting independent from the use of cars. Half of the users (46 percent) avoid a car trip by using a shared cargo

FACTOR 8

bike. An increasing number of European cities (such as Grenoble, Strasbourg, Hamburg and Stuttgart) are integrating cargo bikes into their conventional bike-sharing fleets. In Switzerland, the commercial cargo bike-sharing system carvelo2go currently runs over 300 e-cargo bikes in more than 70 cities. In Brussels the Cairgo project aims at implementing cargo bikes by improving the whole ecosystem, for example by providing more parking spots for cargo bikes.

For the railways, there are still many bottlenecks to the smooth handling of goods and to obtaining information on the whereabouts of these goods in Europe. Rail is particularly interesting for longer distances, from 500 km.

Finally, a shift in the transport of goods is also possible in terms of the time of delivery. Night distribution makes it possible to avoid traffic jams and use the capacity of a vehicle fleet more optimally. It also brings benefits in terms of air emissions and climate. CO_2 emissions can be up to three times lower by switching from day to night deliveries. This depends very much on the extent to which the daytime transport is confronted with congestion. In congestion, a vehicle and certainly a heavy truck, consume relatively more than in free flow. In addition to this reduction in emissions, the time savings achieved by driving at night also provide opportunities to improve the efficiency of delivery rounds. For example, extra stops can be delivered, which means that an additional vehicle that would have been used during the day can be saved. This results in additional emission reductions[111]. Ensuring that this happens in a quiet way can be done by adapting the material, but also by having the trucks drive electrically.

There is still a huge potential to shift to inland waterways and rail. A 2019 study shows that there is still potential especially for bulk and palletised goods[112]. The potential for these cargo types increases even more if the cost of road transport increases by 10%[113]. Congestion or lack of capacity is not an issue on inland waterways. However, both inland navigation and rail should be given more support. For example, in the port of Antwerp, inland vessels are treated as subordinate to the large maritime vessels. Smaller barges are even refused access. They have to carry a minimum number of containers. The railways, too, still have a lot of interoperability problems because the railways between the different countries are not well coordinated and a lot of time is lost at border crossings.

Nonetheless, the larger volumes that can be transported by ship or train do mean that, from certain distances, the prices are better, and, above all, there are fewer emissions per tonne of goods transported. More information about the goods, but also about the capacity of the trains and ships will lead to more use being made of them, so-called synchromodal transport. This involves the use of real-time information to determine which route and means of transport is the most optimal at that moment[114].

FACTOR

ANTICIPATION OF NEW TECHNOLOGIES

The fourth A stands for Anticipation, primarily regarding new vehicle technologies. If we cannot avoid travel, or shift to more environmentally friendly modes of transport, can we at least "change" the cars we use? Here, indeed, many new options have emerged in addition to the classic petrol or diesel cars. As indicated in the 'Awareness' section, electric cars are the most appropriate in terms of CO_2 emissions. They are also a good alternative in terms of local emissions, since there are no more emissions while driving. As with other cars, there is particulate matter from brake disc wear, but this is less with electric cars because they are automatically braked by the motor, so the brake discs wear less than with other types of cars.

If we want to achieve the climate objectives, the electrification of the entire vehicle fleet is necessary. However, we absolutely must also focus on the two previous As, Avoidance and Acting and shifting, otherwise we will not achieve the climate objectives and we will continue to be stuck in traffic jams. Hence, we must aim for less driving and the sharing of cars so that less has to be produced. This is once again confirmed by a 2019 study by VITO and the Circular Economy Support Centre which showed that if we continue to have the same size car fleet, we will not be able to meet the climate objective in 2030[115].

Some countries have already indicated that they will no longer sell cars that are not emission-free. In the Netherlands that is in 2030, and in Norway already in 2025. Cities are also starting to indicate that diesel cars, and later petrol cars, will no longer be welcome; in Brussels this will be in 2030 and 2035 respectively. Such measures are important because a car is used for an average of 15 years and will therefore continue to pollute for a long time once it is bought. If we want to have a completely climate-neutral transport system by 2050, then this is only possible through such clear measures. And to me, the only clear measure is moving towards electric cars. The car lobby has long insisted on keeping such decisions and measures technology neutral. The technology of electric cars could not be chosen decisively. However, none of the other technologies can achieve the climate objectives. Running cars on natural gas is not the solution because it still involves a lot of CO_2 emissions since it is also a fossil gas. Biofuels are not sustainable enough to mix in large quantities. What's more, it will mean that even more forests will have to make way for farmland, taking even less CO_2 out of the air. And hybrid cars are not the long-term solution either. They only run on electricity for a limited amount of time and from experience I can say that the need to switch them on after use feels just a little less urgent than with a full battery electric vehicle which means even less driving on electricity.

However, there are still some barriers to the full market introduction of electric cars. These are mainly the purchase cost, the limited driving range of the cars and the charging infrastructure that is not yet available everywhere. We will see that these barriers are also fading.

FACTOR 8

First, let's look at the cost of the car. It is best to look at it over the entire life cycle of the car in a so-called Total Cost of Ownership (TCO) analysis. In such a TCO, all costs are included - the purchase cost, but also insurance, fuel costs, and so on. For smaller city cars, electric vehicles are currently more expensive than petrol cars. This is also a very competitive segment where you can buy a petrol car for as little as 10,000 euros. But even in the middle class, electric cars are not yet competitive with traditional cars. It only becomes more compelling in the premium segment, for example Tesla[116].

We expect this cost comparison to change soon. This is because the cost of the battery, an important element in this comparison, is rapidly becoming cheaper. Between 2010 and 2018, the cost has already dropped by 80% and is further decreasing due to increased battery production capacity. In 2022, the cost has already dropped by 89% and there is forecast of a further 92.5% in the 2030s[117.]

This is accompanied by increased capacity for the batteries. They are becoming more efficient and are also increasing the driving range of the cars. Electric cars can now easily cover 350 to 400 km with some models going above 600 km, whereas the first models had to do with 100 km. See the figure on page 81.

According to a study by Bloomberg[118], 2026 will be the tipping point for the purchase price of electric vehicles. At that point electric models will be more interesting than cars with a fuel engine. Other studies speak of 2022 as the tipping point. As indicated above, there are already segments in which it is more interesting. A fiscal policy aimed at stimulating electric vehicles can also help a lot. In Norway, for example, 1 in 2 cars are already electric, partly due to a tax policy that makes electric cars cheaper than diesel or petrol cars.

Moreover, there will be many new models on the market by 2022[119]. So, for those who think "all well and good, but those electric cars look so ugly". Well, according to an IEA report, the number of electric vehicle models was 370 as at 2020 and many more models are in production. So, you have several options but if you still can't find your choice in

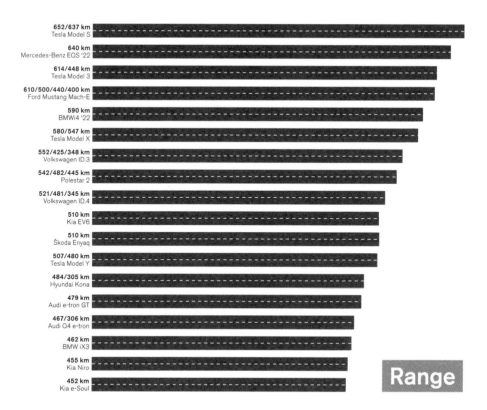

	Range
652/637 km Tesla Model S	
640 km Mercedes-Benz EQS '22	
614/448 km Tesla Model 3	
610/500/440/400 km Ford Mustang Mach-E	
590 km BMWi4 '22	
580/547 km Tesla Model X	
552/425/348 km Volkswagen ID.3	
542/482/445 km Polestar 2	
521/481/345 km Volkswagen ID.4	
510 km Kia EV6	
510 km Škoda Enyaq	
507/480 km Tesla Model Y	
484/305 km Hyundai Kona	
479 km Audi e-tron GT	
467/306 km Audi Q4 e-tron	
462 km BMW iX3	
455 km Kia Niro	
452 km Kia e-Soul	

these, I don't know where else you would. Even Harley Davidson now has an electric model. The salespeople still hope that there will be an appropriate sound but what is disappointing in this whole transition is that the car manufacturers give priority to their big SUV-type cars to bring on the market. This means that more and more big cars are driving around, a trend that was already going on before with the petrol and diesel cars. Now, 45% of global car sales are SUV's [120].

I used to have an electric car at home, which I used from time to time. The car, a BMWi3 was still an 'old' model with a range of 100 km. It had a generator that would create some extra electricity anytime the battery ran out. They call it a range extender. That has helped me a lot because with a range of 100 km, I can only make the journey from Brussels to Antwerp and back, if there are not many detours. But the car drives wonderfully. It is very nimble too, which gives me a safe feeling, because

FACTOR

in dangerous situations I can accelerate very easily. At the moment, I do not have a car anymore, as I am comfortable with the Avoidance and Act and shift part of this book, but for sure I understand that people can be car dependent, and so an electric car can be a solution. Still, it is also good to see if you can share one.

A third barrier for the introduction of electric cars is the number of charging stations in public spaces. People are often anxious about the possibility of their battery running out while they are still on the road. In most cases people will charge either at home (45% of all charging) or at work (also 45%). The other 10% of charging is on the road (5%) or at the destination (5%). People with an electric car will therefore have to be able to charge in those places. Hence, charging infrastructure, and especially fast chargers, are especially important for very long distances where you can refuel in between. In Europe, there are currently 34.000 fast charging points.

People with a garage can easily charge at home[121] and companies are also starting to provide charging infrastructure. However, people without a garage will also need a charging infrastructure. In Brussels, for example, only about 10% of homeowners have their own garage, in Flanders this is more than half. For people without their own parking garage, charging stations will therefore have to be installed in the public space. In many countries, the principle of 'station follows car' is used. In Flanders and Brussels, you have a 'right' to a charging station. Once you can prove that you have bought an electric car, you are entitled to a charging station near your home. However, this is a public charging station, so the use is shared. To prevent people from continuing to occupy the parking space when the car is already fully charged, a double charge is now often applied, which encourages you to free up the space for others after charging[122]. In addition to public charging points, companies like SparkCharge, a US start-up, will soon make charging even easier through its development of mobile electric charging infrastructure that electric car owners can take with them anywhere they are. However, this innovative technology is still geographically limited in distribution[123].

A question often asked is: will we have enough electricity for all those electric cars in our fleet?

If you look at it at a macro level, completely replacing our car fleet with an electric car fleet will only require 20% more electricity. And, of course, this will not happen overnight, the transition will be gradual[124].

But this is absolutely feasible. Indeed, investments will have to be made to gradually provide more capacity. And as we saw, the more the electricity is generated from renewable energy, the better the impact on the climate. So, this transition goes hand in hand. At the local level, of course, shortages may occur if, for example, all the people in a neighbourhood start charging together when they get home in the evening. But that problem should not be exaggerated either. A transition is also taking place in the energy sector. We are moving from a centralised system towards a local and decentralised energy system, in which citizens are no longer just consumers, but also producers thanks to solar panels or even the joint purchase of a wind turbine. Thus, they become prosumers. This has many advantages, especially in combination with the use of electric cars. The batteries in the cars or even in the house (possibly as a second life for a battery that no longer has its full capacity) can be used as a buffer for the energy that is generated and not consumed directly. For example, a company with a lot of solar panels on the roof could charge its employees' cars during the day when the sun is shining and thus store solar energy optimally. Those employees then go home and can use that electricity during the 'evening peak' for cooking and the likes. We call this 'Vehicle to Home' (V2H) - the car's battery supplies energy to the home. Charging then becomes a smart thing thanks to *smart metering* but also thanks to optimisation systems that suggest the best way to charge to the user. This way, a vehicle does not always need to be charged quickly. After all, that would demand a lot from the electricity grid at that particular moment. Charging slowly is often better and this way, the load in a neighbourhood can be greatly reduced.[125]

As you can see, the introduction of electric vehicles, together with the transition in the energy sector, is an optimal combination that will provide an important piece of the puzzle. It will thus be a different capacity of electricity supply and especially a different use of capacity, spread over time, that will be necessary. Therefore, the use of electric cars brings with it new opportunities, because they can be used as an additional storage buffer.

Autonomous cars[126]

What about the future? The emergence of autonomous vehicles is the next hot topic in the development of transport technology. For years, Google, TESLA and others have been testing fully autonomous vehicles that can drive without the intervention of a human driver. The automation of vehicles has various levels, ranging from level 0 where there is no automation at all to level 5. Many cars already contain some automation, such as advanced cruise control or sensors that warn you if you leave your lane. At level 5, there is no longer any need for human interaction with the vehicle. Higher levels mean more automated functions and less human intervention, but also more technological complexity. According to the industry, they are ready to introduce fully self-driving vehicles and they should become part of the regular street scene within 10 years. By 2030, autonomous cars with human assistance (a person will still need to be behind the wheel) are likely to be commercially available[127]. Ten to twelve years later, driverless autonomous sharing systems may be introduced in cities[128]. However, experts are cautious and indicate that there are still some questions and bottlenecks that need to be resolved first.

The question of how autonomous vehicles will change our lives is difficult to answer. I like to use Albrecht Dürer's woodcut of a rhino (1515) to illustrate why it is not so easy to imagine:

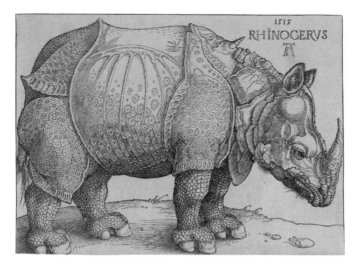

Dürer himself had never seen a rhino. Based on descriptions and a sketch by an unknown artist, he made this creditable work of art, which was copied many times afterwards.

Several of these pilot projects are now underway[130]. But of course, such pilot projects do not tell the whole story and only reflect a fraction of what the effect of an autonomous vehicle will be on our lives. A nice (but rather expensive) experiment to check what the effect would be on travel behaviour was carried out in the US. In order to simulate autonomous driving, 13 families were given a car with a driver at their disposal day and night, to see whether they drove more or not. And what do you think? Yes, the number of kilometres driven increased by 83%, especially in the evening hours and often to drive around other people like the children. Getting the children to their hobbies was suddenly no longer such a problem[131].

So, it looks like it will generate more vehicle kilometres. This is quite logical: after all, the time lost on the journey is reduced. If you can work or play on your smartphone in the meantime, you won't mind being in the car as much. Moreover, more people can get into the self-driving car. Young people without a driving licence, people who cannot drive because of a disability, older people who have given up driving - they are all included. Autonomous mobility means more freedom for these population groups. Most studies predict an increase in vehicle kilometres of between 40 and 80%[132]

A very important question will be whether these autonomous cars will be shared cars, integrated with public transport, or individually owned by their passenger(s). In the latter case, we can expect even more congestion. It may be felt less harsh because in the traffic jam you can still be busy with all sorts of other things, but it is not a solution for a sustainable mobility system.

If the cars are shared and also connected, the positive effects are much greater[133]. Connected autonomous vehicles are cars that are connected to other cars (directly or via a control centre) and even to the road infrastructure. Autonomous connected buses can then immediately incorporate calls from waiting passengers into their flexible routing and dynamic timetable, even during the journey, which improves the service considerably.

If they are shared, we need far fewer cars. A study by Tractebel has calculated that 21,000 autonomous shared taxis would be enough to cover all journeys in Brussels, compared with the 330,000 vehicles and 621 buses that currently make these journeys. Only 62,000 shared cars would be needed to travel to and from Brussels, compared to 400,000 cars today. A simulation study for the Lisbon region by the International Transport Forum reaches the same conclusion[134]. If we start sharing vehicles en masse and combine them with the public transport network, congestion can be solved and there will be a big improvement in local and global emissions. This assumes that a trip can easily be 'ordered' via, for example, an app and that a vehicle will then quickly arrive near the place where you requested the trip - more like the current Uber service but without a driver. For the users, the comfort remains very high: the trips are often made without changing buses and are geared to good connection possibilities with public transport so the costs are halved, which contributes to more equality for all mobility users. In such a partial scenario, the purchase price of a vehicle is no longer important, but transport is charged according to use. One 2019 study goes further in its analysis for the best implementation scenario for autonomous vehicles[135]. In the best scenario, there can be a reduction of vehicle kilometres by 14 to 31%. This assumes that people who currently use public transport continue to do so and that people who currently own a car start sharing trips. In the worst case however, we have a doubling of the number of vehicle kilometres if people who now take the bus and tram also start using shared cars (without sharing rides). Since the assumption is that people who take the train and metro will continue to do so, the anticipated progress can ultimately be questioned.

It will therefore depend very much on how autonomous vehicles are given a place in our mobility system. If they are indeed shared and connected, there will also have a very positive impact on the use of space in cities. Shared autonomous transport systems can make 90% of current parking spaces in cities redundant, which means you can do very nice things with the freed-up space[136]. In addition, the capacity of the current road network will be better utilised through a more efficient use of the road surface, as vehicles will drive more densely and less unpredictably. The road infrastructure can then be used more efficiently (for instance, four lanes to drive on during the evening rush hour and only one in the morning).

San Francisco Smart City Project[137]

FACTOR 8

Road safety is also expected to improve significantly. 72% of all accidents are due to human behaviour[138]. An autonomous vehicle is much safer and cannot make those human mistakes. However, other ethical issues arise. How should the algorithm be set up: does it save the child crossing the road or does the algorithm choose to save the driver?

Not all effects will be positive. For example, commuters may sleep or work in their car while travelling to work. This may motivate them to live even further away from the cities, further increasing fragmentation and having a negative impact on land use. It is also unclear what the impact of automation on employment will be when bus, taxi and truck drivers are no longer needed. This can lead to great social resistance to automation. But in every transition, you get such shifts. In the transport sector, this is not really seen as a problem because they have been struggling with staff shortages for years.

So, we are hearing a lot of wild stories about the rhino/autonomous car, and we may have seen some rare specimens, but we can only speculate about a full fleet that will potentially influence our mobility behaviour very strongly. What is clear is that the advent of the self-driving car will change not only our travel behaviour but also our way of life. If we want this robotisation of the transport sector to evolve into a positive revolution for the benefit of all (urban and rural residents, mobility users, etc.), it is crucial to create the context now with policy measures so that the entire mobility system is ready for such a positive (r)evolution. Such measures include stimulating shared use of cars and shared transport, creating clear frameworks for mobility providers, stimulating innovation and new business models, and protecting users. Europe has reacted ex-post to the arrival of Uber in its cities, and this mobility service is now so well-established that it is impossible to imagine our cities without it. Thus, when a manufacturer/technology giant arrives in a city with a fleet of self-driving cars that meet transport needs in a very user-friendly and on-demand manner, even without the city having prepared on all fronts, the opportunities for a more sustainable transport system is obvious. However, this is only true if it is implemented in the right way. So, let's not wait for the arrival of the rhino's (that is, autonomous vehicles) before have a clear vision on how to implement these.

Anticipation in aviation

As said in the beginning, flying is the most impactful transportation mode. Not only because of the amount of kerosene that is burned, but also because of the radiation forces which further increases the global warming.

Just like in the road transport sector, the industry is aiming for fuel efficiency gains of around 1.5 percent by improving the existing technology. However, with the estimated annual growth rates of about 4 percent, efficiency savings are overall negligible.

There are high expectations of synthetic e-fuels made from electricity. However, these are still very expensive - three to four times more expensive than conventional kerosene. And while this is technically feasible, it would mean that all existing renewable electricity supply in the world would be taken up by the aviation sector. Advanced biofuels (not based on agricultural crops) could also be an intermediate solution. The cost difference is still a major barrier to both production and use. Palm oil would be the cheapest source but that would mean biodiversity loss and pressure on the food market, as previously explained.

Experiments with electric aircraft are also under way. In Canada, an electric aircraft was tested, and it flew for 15 minutes over 160 km. Many of Harbour Air's commercial flights, which take people from Vancouver to ski resorts and nearby islands, could be operated with this[139]. However, as batteries weigh too much, it is not an instant solution for commercial flights.

Still, we are far away from having a sustainable solution for flights. After a short break during COVID people have returned back to flying for their holidays. Weekly flight updates from EUROCONTROL indicates that the average number of flights for the 2022 summer holiday period is at approximately 30,000 flights per day as opposed to the 11,500 of 2020[140].

However, another part of the population has (re)discovered the train for going on holidays. This summer I took the train to go to Malaga. From my departure in Brussels, it took me 27,5 hours and six trains, including a night train from Paris to Toulouse. I liked it a lot as I had a great time.

FACTOR 8

I read a lot, had nice lunches and slept well. On my way back, it took me a bit longer as there was a strike in France and I ended up spending the first day in Figueras, which was a nice discovery. Of course, I could only enjoy this because I was not so much in the hurry.

There is still a lot of work on communication towards the traveller and certainly there is a lack of information sharing between the different countries even if they collaborate for certain high-speed trains. It also cost me more - 385 euro on my way there, compared to a ticket by air of around 200 euro in the big season. But the main benefit is to have had some 7 times less CO_2 impact than going by plane.

Changing the goods transport

In freight transport, it is also important to move towards electrification as much as possible. However, this is not so simple. The heavy battery that would be required is not only expensive, but also takes up a large part of the lorry's authorised mass.

The total cost of ownership which analyses how much the vehicle costs over its full lifetime looks completely different here. In goods transport, the total cost of ownership for small and medium vans is getting in favour of electric variant instead of the diesel one. And so will the larger vans by 2026. For heavy duty vehicles, it will take longer. The larger the freight vehicle, the more expensive the electric variant. But as with passenger cars, price differences will disappear because battery prices are falling rapidly and because freight vehicles too are going into mass production. There are a few examples that show that demand is rising: for example, Amazon is going to put 100,000 electric delivery vehicles into service, DHL has developed its own electric vehicle, the streetscooter, of which there are already 10,000 on German roads and the use of which will be further rolled out in the coming years, and Ikea is working towards electrifying its last mile delivery[141].

Further uptake of electric vehicles will accelerate if cities are to meet the target of making urban distribution emission-free by 2030.

For the larger trucks it is, as said, a bit more difficult, but a lot of new models will be brought on the market. Renault, for example, now have a line-up of electric trucks[142], and Tesla is bringing out an electric model with its Semi, of which they want to build 100,000 a year[143]. These cars can be driven for 6 hours and then recharged.

In order to need less battery, you could also work with a kind of trolley system but for trucks, namely e-highways. It looks like this: In this way, the truck can be hooked up for a short while and does not need to use any energy from the battery. This could be an interesting system for the busy port routes like that of Antwerp. It is already on trial in Sweden, Germany and the US[144].

In addition to electrification, CNG/LNG, biomethane or hydrogen are also being considered for freight transport. Hydrogen may have a role to play in transport over longer distances, but that means rolling out hydrogen charging stations all over Europe. In Belgium, the first has now been opened by Dats21. However, energy efficiency is also a factor here, and it will therefore be a more expensive solution (hydrogen is three times less energy-efficient than an electric vehicle). An interesting combination has been developed by VDL, in which an electric truck is fitted with a range extender based on hydrogen. Such a range extender ensures that if the battery is empty, energy is still replenished - via the extra energy source - in the battery[145].

FACTOR 8

CNG/LNG is compressed natural gas in gaseous form (CNG) or in liquid form (LNG). For small to medium-sized cars, CNG is often used, for larger trucks but also for inland navigation vessels, the switch is usually made to LNG. However, as already mentioned, natural gas is also a fossil fuel. In terms of CO_2 emissions, you only gain 6 to 7%. For local emissions and air quality, there will be a noticeable improvement, but it will not solve the climate problem.

Biomethane also exists in gas or liquid form. Recall that biomethane is made from waste or possibly from crops. So, this could possibly be climate-neutral. However, the chance of methane leaks is real and, since methane is 25 times worse for the climate than CO_2, the risk should not be underestimated. Moreover, biomethane cannot be produced on a large scale and is therefore of little use to the transport sector.

The provision of sufficient charging infrastructure is even more important here than in passenger transport, where there will be more charging at home or at work. Rapid chargers that can charge delivery vans and heavier trucks during their journey or round trip are essential here, especially for longer journeys. Urban distribution can be made emission-free more easily. This involves shorter distances and with the use of urban distribution centres, the goods can be transferred to cargo bikes and electric delivery vans. The electric delivery vehicles already have an interesting total cost of ownership compared to diesel vehicles[146].

Self-driving cars are also being considered in this sector. More so because the sector has a lot of problems finding personnel. According to McKinsey, by 2025, 80% of all packages will be delivered by self-driving freight vehicles[147]. The strangest types of vehicles are being designed for this purpose.

Small robots that can carry three to four packages and drive to the door via the footpath are also being tested. It is expected that delivering goods with autonomous vehicles can reduce the price of transport by 40%[148]. Other atypical things are also being tested, such as Robomart, a self-driving supermarket that comes to you so you can buy fresh products at the door.[149]

Robomart grocery car

The Amazon Scout and the Starship

When we talk about futuristic things, drones should also be mentioned. And then again, not. Drones receive a lot of attention; there have even been tests for transporting one or two people with drones. For goods transport, it seems interesting to have drones deliver packages instead of vans. In terms of environmental impact, the comparison can be interesting, since drones fly on batteries and most delivery vans still run on diesel. However, compared to an electric delivery truck or a cargo bicycle, drones use much more energy. Moreover, for the same volume of parcels as in a van, you need a whole fleet of drones. Can you see it happening? With noise, possible collisions, privacy problems... as a result. In some cities in China and India, drone deliveries have already been tested, and in the US, Google has also received permission. In Belgium, legislation states that in zones where drones are allowed to fly (and this is limited based on the flight paths of ordinary planes), you must always be able to keep an eye on your drone[150]. So, it is better to deliver the goods at ground level. I do not see drones as a solution to the climate problem but there are interesting applications. For example, for locations that are difficult to reach and for fast transport of blood and organs, it could be an efficient means. In other scenarios, not so much.

FACTOR 8

"It would take about 15 drones operating around the clock to deliver the same product volume as a single light commercial vehicle does in a typical 8-hour shift"[151]

But anyway, we were talking about self-driving cars. A first step in this direction is *platooning*, in which lorries drive behind each other – coupled as it were and only the first driver needs to pay attention. Platooning offers some advantages in terms of consumption and thus emissions. There are also opportunities for change in inland navigation. In inland navigation, there are already some examples of ships on LNG, hybrid ships but also fully electric ones. New investments in inland navigation are often difficult because in many cases these are family businesses with little capital behind them. So, there are certainly many opportunities for greening the fleet in goods transport too, even if it is not as simple as in passenger transport. Here we will have to look much more closely at what the new technologies make possible. Nevertheless, investing in greener vehicle technology is crucial for logistics. In our study for the Flemish government, we see that between 67% and 88% of the emission reduction will have to come from greening the fleet. Thus, "Avoidance and Shifting" is important for this sector, because without shifting, we are nowhere.[152]

ACCELERATE

So, a transition is possible: Avoiding, Act and Shifting, and Anticipation are necessary to attain a more sustainable mobility system. Awareness is necessary to make us willing to go for this change! But how fast will this transition to a more sustainable mobility system take place? And will we be in time to achieve the climate objectives?

Some major transitions have happened very quickly; think of the transition from horse-drawn carriages to cars. In these beautiful photographs of 5th Avenue in New York, we see that in 1900, there is only one car among all the horse-drawn carriages; 13 years later, the situation is reversed! A funny anecdote from that time was that people didn't know what to do with the horse-drawn carriages and especially with the excrement of the horses. So, not only were they huge tonnages, but their excrements were also breeding grounds for diseases. A congress was held to discuss these horse droppings; thereafter, the arrival of cars solved that problem but brought about many other problems that we now have to solve.

Easter 1900, 5th Avenue New York,
spot the automobile

Easter 1913, 5th Avenue New York,
spot the horse

So, the current transition we are aiming for can also go fast. In fact, many experts are comparing the introduction of electric cars in the fleet with this previous "horse-drawn carriage to cars" transition. However, you should know that the whole turnaround took about 50 years and was not as easy as these pictures suggest because in any transition there are always winners and losers with the losers trying everything they can to maintain the status quo[153]. Hoping that this transition will happen on its own seems too dangerous especially because we do not have much time left. By 2050 we need to have reduced CO_2 emissions by a factor of 8 and by 2030 (which is already less than 10 years away!) a reduction of at least 55% needs to be achieved. The corona crisis has taught us a lot; and hopefully we have also learned from it that sometimes, drastic measures have to be introduced to contain a crisis *(I wrote more about this in the epilogue to this book).*

If we want to facilitate a move towards a sustainable mobility system, we need to stop hoping that technology will solve everything or that consumers will make the right choices based on sustainability information. Rather, we need the right policy framework that will ensure that sustainable options become the most logical choices for everyone and not just for those who like to keep up with new sustainable trends. Unfortunately, that right framework is currently not in place. At the European level, goals are stated many times and directions are given, for example already in the 2011 White paper (Roadmap to a Single European Transport Area), but there is a lack of implementation at the member states level. There are three major problems (and many smaller ones, but let's look at the major faults of the system).

First, the tax system for cars in many countries is based on a fixed car tax at the time of the car purchase (registration tax) and an annual amount irrespective of how much you drive (the road tax) but this does not encourage you to drive less. On the contrary, since you have paid, you will try to use the car as much as possible. You could say that excise duties on fuel are a form of road pricing (or kilometre charge), because the more you drive, the more you pay. However, excise duties do not cover the full cost to society. Moreover, a diesel car will pay less excise duty because of its more economical use, but diesel cars will emit more fine

FACTOR 8

dust and nitrogen. Although road tax and other possible levies, like the péages (toll) in France, mean that drivers already pay a lot more; we are still a long way from a correct price.

This can be seen in the figure below. The bars on the left show the cost of various external effects (such as congestion, accidents, emissions...) to society while the bars on the right shows how much is paid in taxes.

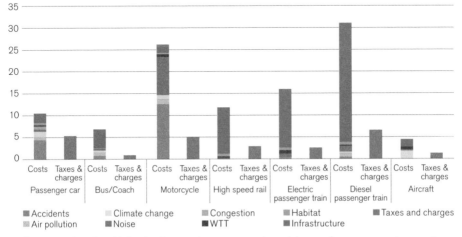

Average external and infrastructure costs vs. average taxes/charges for passenger transport

Calculations of external costs thus indicate which costs are not included in the price of transport but are borne by the society. In economic theory, this gives rise to excessive demand since the price is not correct. Thus, internalising the external costs, i.e., ensuring that the true price is paid, can correct this imbalance.

Second, in many countries, the government subsidises car driving by providing a favourable tax system for company cars. The company car system gives the government less income if we consider the taxes that do not have to be paid. In Europe, newly registered company cars in 2019 accounted for 57% of all cars sold[154]. The fiscal benefits accorded to company cars amounts to about 32 billion euros per year and this cost is borne by the taxpayers. Clearly, this has so many negative effects on the society.

Research has shown that people with a company car travel more kilometres every day (see figure below) and are less likely to use public transport.

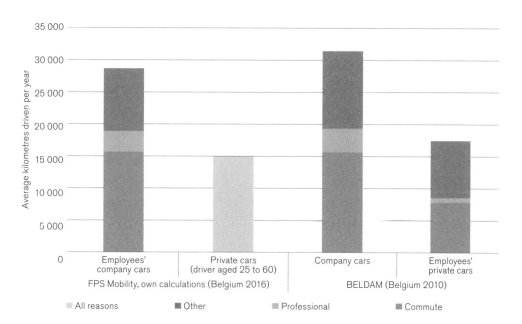

In the figure we can indeed see (based on two different sources) that there is a large difference in the number of kilometres travelled and that only a fraction of this can be explained by driving more professional kilometres (such as for people who use the car not only for commuting but also for their job). Moreover, the vast majority of company cars are diesel cars, while private cars are more often petrol cars. So there is also a large health cost associated with using company cars, on top of congestion and CO_2 emissions.

A third major problem is that the government is not investing enough in the alternatives. The investments in active modes of transport (walking and cycling), public transport infrastructure and other motorised mode is shown in the graph below. It shows the modal split in several cities in Europe and the area covered by the public transport authorities (PTA).

101

	main cities			PTA areas		
	% active	% public transport	% rest	% active	% public transport	% rest
Amsterdam	55	14	31	42	9	49
Barcelona	67	17	16	36	11	53
Belgrade	25	50	25	25	50	25
Berlin	42	25	33	44	19	37
Bilbao	66	22	12	35	18	47
Birmingham				70	10	20
Budapest	18	47	35	35	47	18
Copenhagen	66	10	24	49	6	45
Frankfurt	48	24	28	55	12	34
Helsinki	44	31	26	40	22	38
Krakow	31	30	39	41	30	30
Lisbon	30	22	48	60	16	24
London	34	27	39	39	27	34
Lyon	48	25	27	44	19	37
Madrid	35	33	32	41	24	34
Mallorca	41	13	46	55	10	35
Manchester	34	9	58	65	9	26
Oslo	43	25	32	48	19	34
Paris	68	26	6	36	22	43
Porto						
Prague	28	43	29			
Rott/Hague	54	3	47	45	3	52
Stockholm	55	22	23	51	18	31
Stuttgart	46	17	37	60	8	32
Thessaloniki	37	24	39			
Turin				62	10	27
Valencia	48	21	31	43	14	43
Vienna	46	27	27			
Vilnius	26	24	50	50	24	26
Warsaw	21	47	32	36	40	25

100% 90% 80% 70% 60% 50% 40% 30% 20% 10% 0%   0% 10% 20% 30% 40% 50% 60% 70% 80% 90% 100%

■ % active modes   ■ % public transport   ■ % rest of motorised modes

EMTA Barometer 2022 (based on 2020 data – 16th Edition[155])

Each city on the graph has its own mobility policy, hence a clear difference in modal split. For example, Barcelona and Copenhagen score very well in the use of active modes; Lyon also score well on active modes primarily because the government has invested heavily in a walkable city without cars; while Prague and Budapest score well on public transport. As far as cycling is concerned, the main examples where the transition has worked are shown in the graph below.

The share of cycling decreased everywhere after the advent of the car and its mass production. However, some cities have refocused their policies on cycling by providing the necessary infrastructure and taking cyclists into account in all decisions. In those cases, transformations are taking place. Copenhagen is the most cited example, and this did not happen overnight but was created by politicians with a long-term view. It has to be safe, fast and easy to get around by bike in the city and then the citizens will be motivated to start cycling. So, cycling transformation in Copenhagen is neither because it has better weather nor does its residents have an extra interest in it. Simply put, a network of protected bike lanes and enough parking infrastructure must be built throughout the city. That being said, these three major underlying problems in the current policy framework need to be addressed. It will be elaborated upon below how this might look like in concrete terms, based on the transition needed to an Avoid, Shift and Change policy.

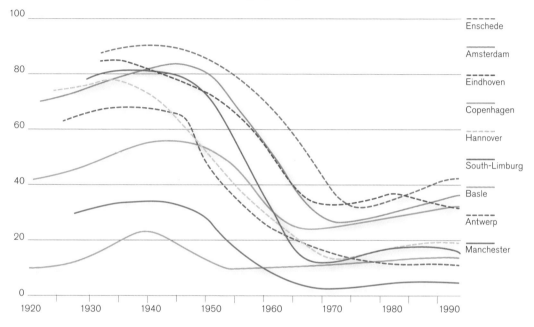

Avoidance

In order to take full advantage of Avoidance, the government should focus on densification and multifunctional developments. The first major condition to achieve a different and more environmentally friendly mobility, is a different organisation of the space through legislation. Spatial planning aimed at densification ensures that there is no further fragmentation, but active intervention could also be undertaken. For example, by adapting the housing tax system; people who live in a detached house in the countryside will see their taxes increase, while those who live in the city centre will see them decrease.

FACTOR

5 ACCELERATE

Of course, this will be a measure that has to be rolled out over a longer period, but for that very reason, it has to start tomorrow or even better today, otherwise valuable time will be lost. If rolled out over a longer period, the necessary compensation and thus, the cost to government and citizens will be more limited, but the benefits will be great for government, citizens and businesses.

Act and Shift

In the Act and Shift phase, several measures are needed together. First, the alternatives must be worked on; and this is shown in the examples above. Subsequently, the impact of travel on society, which includes impact on the climate, air quality, accidents, and so on, should be included in the price of travel. A smart road pricing system, which incorporates all these external costs, makes this possible.

Taking Belgium as a case; for the transport of goods by road, the country already has a distance-based charge for trucks over 3.5 tonnes. The charge considers the environmental friendliness of the truck (based on Euro standards) and the location where it is driven. For example, lorries have to pay more if they drive in the Brussels Capital Region. Even so, the kilometre charge only applies to the main road network and some adjustments have been necessary to prevent cut-through traffic. The kilometre charge for trucks does not consider the difference between off-peak and peak hours, which means that trucks are not encouraged to avoid the traffic jams.

Interestingly, the extension of the smart road user charge to passenger transport has already caused much discussion and has led to many research studies. According to research for the Flemish government, such a smart, area-wide road user charge system where drivers pay a charge per kilometre driven and where the rates vary according to time of day, place, road type, direction of travel and/or vehicle characteristics, could achieve the objectives of reducing congestion and emissions. Moreover, it would bring about a system where the polluter pays, whereas under the current system everyone pays taxes for the maintenance and construction of roads, even if they do not use them.

The underlying idea of road pricing is very good - namely he who pollutes more, pays more. The level of pollution can be determined by knowing the type of car (how much air pollutants and CO_2 are emitted), where the car is driven (if many or few people live there to determine if the effect of these emissions is greater or lesser) and at what time of day the car is driven (to estimate the level of congestion). The result of implementing this idea would be a reduction in the number of journeys and a shift towards public transport and cycling. But there is a lot of opposition to the system because people think they will have to pay more. Indeed, some will have to pay more, but others who do not drive much but currently pay a fixed annual amount for road tax will pay less.

Under the new tax system, a study for Flanders calculated that car drivers who drive up to about 17,000 kilometres a year would gain on average, especially if it is also accompanied by a tax shift, but how much they will gain will depend on the type of car they have and where they drive. However, the 40% of households that drive more are worse off. Most households that drive more than 17,000 km/year, and all households that drive more than 31,000 km/year, will experience a decrease in their purchasing power if their behaviour remains unchanged[156]. For example, someone who lives in the countryside and drives mostly outside rush hours with a relatively recent vehicle with Euro standard 5 and drives about 15,000 km per year would probably pay less under the new system while someone who has to drive mainly in the congestion zone with a similar profile will probably pay more, but they can easily find other alternatives in that zone. Hence, providing the population with good examples and further information will help dispel their fears.

Cases from countries where the system has already been implemented show that there is always considerable resistance at first, but that people become very satisfied with the system afterwards. After all, it brings many benefits in terms of congestion and the revenues can be used to invest in environmentally friendly options.

In Europe, all the examples of smart road pricing for passenger transport are actually tolls[157]. In London, Stockholm, Gothenburg, Milan, Oslo and a few other cities, there is a fixed amount that is paid to enter the city or

FACTOR 8

the wider region. The introduction of the congestion charge in London, which is currently at GBP 15/day or about EUR 17.65/day, has reduced congestion by 30% and the number of private vehicles entering the charging zone by 21%. Under the Area C scheme in Milan, entrance is either denied, free or charged 5 EUR/day depending on the type of car and fuel used by the cars driven into the area. Even residents of the area are charged 2 EUR/day if their car types are not among those eligible for free entry[158]. As a result, the scheme has increased bus speeds by 6.9% and increased tram speeds by 4.1% during peak hours.

If many other cities in Europe were to introduce a toll, this would also have positive effects on the purchasing behaviour. For instance, if a differentiation is made according to the type of car and electric cars are allowed free entrance, there will be a major leverage for the purchase of electric cars. This was the case in Milan and Oslo where the fee for electric was free while others paid. However, electric cars drivers in Oslo now have to pay because they also cause congestion and take space.

Moreover, the revenues from the toll and smart road pricing allow for further investment in the alternatives or for a return of the money to the citizens through other means, such as personal income tax or reducing the burden on labour.

Another reason for governments to think about the introduction of a road pricing scheme is that the transition to electric cars will eliminate the income from excise duties, so, a different tax system needs to be used to steer them.

Proactivity but also regulation is needed for new forms of mobility (speed-pedelecs, micro-mobility, and so on). Shared (electric) mobility, MaaS and mobility hubs are developments in which the government must stimulate and facilitate the market. Shared mobility is, as indicated, an important step in the future mobility evolution. Together with MaaS, it can reduce car ownership. Once people no longer have a car, the choice for public transport, cycling and other options is a much more logical step. Investing in reducing car ownership is therefore very important. This can be done by not abolishing the tax on traffic but, on the contrary, by maintaining it and making it even more manageable. The switch to partial systems and MaaS and getting rid of the car can be stimulated by campaigns that encourage trying out these services and by subsidising them in the start-up phase. For MaaS, a guiding framework must be worked out; it must be made clear who takes the lead. Fares and ticket integration between public transportation companies are matters that should already be introduced, even without the idea of MaaS.

Investing in alternatives may sometimes seem like an expensive venture but some measures don't cost that much. For instance, improving bike use can be achieved by simply including footrests at intersections, tilted garbage cans or making bicycle air pumps available in urban space. Whether it is simple things like basic bicycle infrastructures, having a reasonably managed taxing system or using a toll to reduce car ownership, the government can certainly be doing more. Nevertheless, the efforts shouldn't be limited to road transport but other modes as well.

For the European rail system, the issue of hidden borders (non-existent or non-operational cross-border rail connections) needs to be addressed. Historically the railways systems were built within each country, and they differ in many aspects such as gauge, signalling, power systems or regulations. Even now many countries still have to be convinced to invest in cross-border infrastructure. This hampers the further development of the European railway system.

There is also a problem in terms of the communication between the different countries. For example, on my way back from Malaga to Brussels, they did not know in Spain that there was a strike in France and as a result, my train trip would stop at the border in Figueras. In such events, they are not willing to take responsibility at all. One day, one ticket for the whole of Europe would be great to have, together with connected train services. Already in some countries interesting tariffs are proposed like the Klimaticket Ö for 3€ a day in the whole of Austria or 1€ per day for one state. Some states, such as Vorarlberg and Vienna, have introduced the 365€-Ticket (1 euro per day) and have seen a clear increase of users of public transport. In Germany they introduced unlimited travelling on the public transport system for 9€ in the summer of 2022. Seventy percent of the people that used it, said they would use public transport more in the future[159].

FACTOR 8

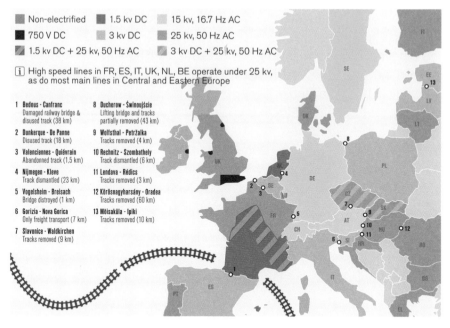

Europe's Hidden Borders[160]
Examples of missing cross-border railway connections in the EU-28 and Switzerland, 2012-2020. Different electrificaton systems and missing cross-border connections complicate rail traffic across borders.

So, there is potential, but we are still far from having a European wide transport infrastructure as promised by the transport network corridors the EU has set up. What is needed is a real integration in terms of ticketing, data and infrastructure. This EU-wide transport system should have as its main objectives climate neutrality, inclusiveness, sufficient space and capacity.

The same observations apply to freight transport. Further investment should be made in the alternatives and the cost price should include the external costs. Although, a recent study by Mobilise shows that it can also lead to additional emissions if all external costs are considered without correction in goods transport. Due to the high congestion costs, a simple introduction of a kilometre charge could lead to detours, which would not be good for CO_2 emissions. What can also happen is if a kilometre charge is introduced for goods transport, lorries will also make detours and the kilometres driven will increase again. Hence, it is important to properly assess the possible negative effects and to coordinate with the surrounding regions[161].

Furthermore, intermodal and synchromodal transport of goods requires organisational support: transshipment facilities, sufficient available network, data, etc. This implies that the port players can also play an important role in stimulating more ambitious modal split targets.

Aviation

As indicated in the introduction, people are flying more and more. Ticket prices have therefore never been so low. What's more, there is no excise duty on kerosene and no value added tax (VAT) on tickets as well. The aviation sector is also only partially part of the ETS system. In 2008, there was a decision to add the sector, but in 2012, just when it was going to enter into force, there was a big protest from the other countries, and they decided to add only the flights within Europe to the ETS system, therefore the impact on ticket prices is very limited. However, flights outside the EU will be covered by CORSIA (Carbon Offsetting and Reduction Scheme for International Aviation). CORSIA is a global carbon offsetting scheme that was developed by the ICAO (International Civil Aviation Organisation). The goal is carbon neutral growth from 2020 onwards. This programme consists of two main elements: modernised and efficient technologies and operations, and carbon offsetting. As explained above, a real technological solution for the aviation sector is highly uncertain, so it means the sector is mostly looking at offsetting. However, this is not without risk as it can go hand in hand with land grabbing and displacements of local communities.

The government could also ban short-haul flights. Some airlines, like Qatar Airlines are flying very short distances (for instance, between Maastricht and Liège to bring cargo from one place to another). There is a growing awareness that the social cost of such flights is too high. Also, celebrity private jets are becoming less and less accepted (you can follow them via the twitter account @CelebJets). For example, an 18-minutes flight from a private jet from the rapper Drake from Hamilton,

FACTOR 8

Ontario, to Toronto, was good for 5 tons of carbon; more than double of the maximum yearly emission a person is supposed to cause if we want to stay within the 1.5°C boundary. The reactions to such behaviour show that there is more awareness, and that this kind of "normality" is increasingly questioned. Passenger flights can just as easily be replaced by train journeys, however, if we really want to accelerate this transition, train ticket prices and comfort should be cheaper so that the sustainable option becomes the logical choice.

Another possibility is to introduce a frequent flyer levy, which would allow a levy-free flight every couple of years, but make every other flight taken more expensive. 80% of the world population is not flying at all. But some are flying a lot and have a huge impact. In the UK, 1% of the residents are responsible for almost 20% of the flights[162].

The aviation subsidies amount to 30 to 40 billion euros annually. And then we should not forget the enormous amount of support the aviation sector got during the COVID-19 crisis. Billions of Euros were given without any meaningful environmental or social conditions attached. If we really want to create a transition, we need to shift the investments towards climate-just sectors such as railway and public transport. Rail infrastructure and services have always had the advantage of long-term reliability as it can last for decades, and it doesn't need as much repairing as the road infrastructures. In the maritime sector, too, there are few binding measures. At the moment, they are just assessing the situation and, above all, holding a lot of meetings. More research into emission-free fuels for these two sectors will be crucial and the real system change will come if we are profoundly thinking of the kind of tourism and travel patterns we want to keep in the future.

Anticipation

As discussed in the Awareness section, the life cycle analysis of cars indicated that battery electric cars have the lowest impact in terms of CO_2. The ultimate goal must therefore be the electrification of the entire vehicle fleet. This can be done in different ways. Some countries, such as France, Norway and the Netherlands, will ban the purchase of cars that are not completely emission-free by 2030. The terminology needs to be carefully considered, by the way. Policy texts often speak of low-carbon and technology-neutral, but low carbon is a misleading term. Emission-free is clear: no emissions. Low carbon can mean anything: a little less bad but still bad. Thus, this is meaningless as a target. Technology neutrality is another term many lobbyists use. It has long been preached in order to let the market play, so to speak, but mainly to preserve the status quo. It should not be said that electric cars will become the norm. That would be too optimistic. However, as a result, we are losing valuable time and people continue to make decisions in an unclear framework or, above all, postponing them.

The decisions of these countries are clear and on a timeframe that also gives car manufacturers the opportunity to adapt. Cities that indicate that no diesel or petrol cars will be allowed in from a certain date onwards, such as in Brussels from 2030 for diesel and 2035 for petrol, are also causing a major turnaround in purchasing behaviour. The low emission zones that were introduced in many European cities, which were often introduced without much opposition (in contrast to the kilometre charge/road pricing that we have been talking about for decades), also have an undeniably positive impact on the switch to a cleaner car fleet.

When switching to a tax system based on use of the car, it could be considered whether the registration tax of cars should be retained. After all, at the time of purchase this tax also has an important effect on choice behaviour. If this tax is linked to the environmental impact of the car, you can again encourage people to buy the least polluting cars. In addition, a tax on car purchase can be justified to pay for the environmental costs of the production and use of space of the cars.

Meanwhile, before a complete change in the tax system is being implemented, the premiums for small and medium-sized zero-emission cars should also be increased. As we saw, this category of cars is still more expensive than petrol and diesel cars. A tipping point is expected to occur in 2026, by 2028 at the latest, but in the meantime an electric vehicle should not be just for the richer population. With increasingly

FACTOR 8

strict standards in various cities regarding low emission zones, there is a growing realisation that diesel cars will not be tolerated for long. Car manufacturers are therefore trying to get rid of these cars quickly and the price difference is becoming even greater. Many countries also started to give purchase premiums for electric bikes, like Luxemburg, Sweden, Norway, etc. Research shows that these grants of 25% of the sales prices can increase the sales with 70% and that half of these new e-trips are substituting a displacement with the car[163].

A major lever that can be used to introduce electric cars is company cars. In the problem definition above, it was already indicated that company cars are a problem for sustainable mobility and the system costs the government a lot of money. In addition, it further widens the income gap between employees.

It accounts for 11.5% of the current car fleet and given all the negative effects, also in terms of congestion, it would be better to abolish the system and replace it with a tax shift where less is deducted from wages. But in the context of this factor of 8 story and the urgency of the climate problem, it is better to use this system to accelerate the transition to electric cars. Firstly, company cars are often more expensive cars hence it is much more interesting in terms of total lifetime costs to switch to electric cars. Moreover, company car owners are often people with their own garage, which makes the introduction easy. Companies are also encouraged to invest in charging infrastructure. In our research on company cars, we found that 71% of company car owners would buy another car or replace their company car with their own if the company car system were abolished. Therefore, it is good to get these people used to an electric vehicle. Company cars are also quickly replaced (on average after 2.3 years[164]), and these cars then end up on the second-hand market and can thus become affordable for people who want to buy an electric car but not at the full purchase price. For all these reasons, and because the measure can be introduced quickly, the system of company cars should be converted to a system where only electric cars are allowed. In the long run, the role of companies in a sustainable mobility policy should be considered, even so, the use of the car should not be encouraged through company cars. This will give an enormous impact in terms of climate and congestion[165].

Other measures

There are still other strange quirks in the current policy. It would take us too far to go into every aspect, but one of them is that of professional diesel. In some European countries, professional diesel users, such as logistics companies, can reclaim from the state a portion of the excise duty they pay on the diesel they fill up. In this way, you are actually rewarding those who consume the most.

Although there has been a proposal to revise energy policies[166], this may not extend to national practices of reclaiming excise duties. Usually, there tend to be drawbacks by the degree of autonomy that EU member states have on their national policies[167]. This negatively affects the efficiency of the EU climate policy and certainly mobility policy as well. Not to mention, there is fragmentation of competence that could prevent the implementation of intra and inter regional projects. After all, mobility is not restricted by borders.

ACTOR INVOLVEMENT

Moving towards a climate-neutral society will involve changes in many ways. The digitalisation of transport might mean that some people get excluded from the new mobility system as they might not own a smart phone, internet or a credit card, which will now become a necessity. In Europe 44% of the people do not have any digital skills and the number of people having access to the mobile internet is only 39%[168]. Aside from this, the transition will also mean a change in workplaces for people working in the car industry, for example, people who have to learn new skills and even shift to more sustainable sectors, like the railway sector and biking sector. Hence, to make this transition a just one is very crucial. At the moment, many workers in the transport sector are already underpaid and have bad social security. So, the new changes must address the social aspects as well. A fair transition framework must take a broader perspective by assessing the impact of different measures on the different groups of people in the society.

So, it is clear we need to consider all the stakeholders involved. The ones with power, who might try to hamper the evolution towards a climate-neutral mobility system, the more vulnerable people and especially the next generation.

The Multi Actor, Multi Criteria Analysis (MAMCA) method, which I developed 20 years ago, makes it possible to present different possible solutions for mobility problems to the different actors involved, the so-called stakeholders. Suppose that in a city, you want to reduce car use; you can draw up different scenarios, such as a toll, a car-free zone and transit parking on the outskirts or a firm parking policy. Other scenarios may emerge when you sit down with the stakeholders. We also ask them what their objectives are and what their priorities are. Shopkeepers, for example, will put forward accessibility and a pleasant shopping environment, while citizens are often concerned about traffic safety and local emissions. Next, the different scenarios are evaluated according to the objectives of these actors, giving a clear picture of the advantages and disadvantages of the different solutions for each group involved. It thus allows participation to take place in a structured manner, where hidden agendas are not possible, but where what is of utmost importance for other actors is understood and can be considered. If different institutional actors are involved, as is often the case for mobility projects that cross regional borders (the regional express network, transit carparks, bike routes, etc.), the competences of the different actors can also be considered. For this, too, we have developed an adapted method so that the various actors can express what is important to them, but where they also know after the analysis which part of the solution lies under their competence (Competitive based Multi Criteria Analysis or COMCA).

In the current institutional context, there is a tendency to optimise mobility at the level of the regions or at an even more local level. This is an obstacle to sustainability because mobility issues, and certainly the environmental component, are usually on a larger scale than the time and space in which the institutions can act and be responsible. However, within large mobility projects or measures, the various institutions do not just have conflicting interests, but also common interests. Nevertheless, due to political strategies and institutional culture, different institutions are often poorly informed about each other's interests, or those of other stakeholders. It is therefore important to bring everyone together to identify those potential blockages, but above all, to point out the win-win situations. In the end, climate is a common interest for all actors, so working towards that can become a binding factor and a basis for co-

FACTOR 8

operation. Over the years we have noticed that the value of the method lies in bringing together the actors and the shared information on goals and impacts makes transformation possible. MAMCA is really an empathy tool by structuring a complex decision problem and letting different groups listen to others[169].

A second way to involve citizens is to include them in measuring and evaluating the effects *(citizen science)*. For example, citizens can be questioned via an application on their smartphone, can help measure speeds or emissions by installing a device at their home or can be asked to grow strawberry plants and so on, to determine the amount of particulate matter. The number of possibilities for actively involving citizens in the research project is large. Projects like D-Noses, "Samen meten" initiatives, Hush city app for noise pollution to mention a few have allowed citizens to participate in air and noise measurements across many countries in Europe[170]. The 'curieuze neuzen' project in Flanders is one of these successful projects. Twenty thousand Flemings participated by measuring the NO2 concentrations in their environment. As a result, there is now much better insight into air quality in Flanders.

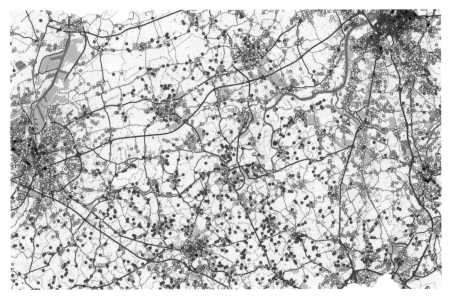

Air pollution concentration in various cities in Flanders

Oftentimes, citizens are willing to collect data. Once they do so, they become much more interested in the subject and even want to be involved in the search for solutions[171]. In Ghent, Belgium, we set up a campaign, in collaboration with local NGOs to measure the amount of time that cyclists loose at red traffic lights. Up to 114 cyclists registered 485 trips via the smartphone app. The study showed that participants on average spent 4% of their trip time waiting at signalised intersections. The results were communicated to the local municipality highlighting specific junctions where the signal patterns could be improved[172].

The board game 'Mobility is a serious game' can also help to get everyone involved[173]. This game, which was developed together with a few people at the initiative of a large retailer Colruyt, makes it possible to explore the future of mobility. The idea behind this development was to make people think out of the box and in a different (playful) way about the future of mobility and logistics by exploring the next 5, 10 and finally 20 years in three game rounds. Participants of the game take the role of a policy maker, company, NGO or even a person who gives the public a voice. All these actors can propose actions, cooperate, invest and then receive comments from that public voice.

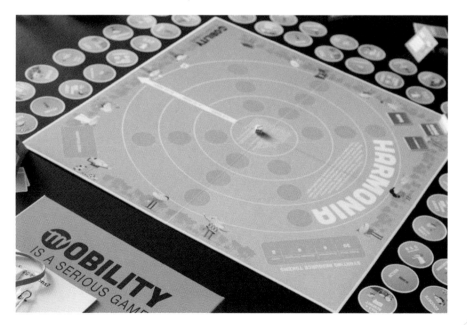

FACTOR 8

It is clear that people change more easily if they feel involved and can form an idea of the consequences of the change. To avoid leaving anyone behind in the transition and to ensure that resistance is reduced, such participatory tools are very important. In addition to the examples given here, participation and co-creation are increasingly in the picture[174], not just because people realise how important it is to have everyone on board, but also because the beginning of change lies within all of us. So, let's explore the steps we need to take to support a change in people's behaviour.

ADAPTING BEHAVIOUR

Finally, we will have to change our behaviour. However, this is easier said than done. In presentations, I always use the joke about my partner, who I have been trying to change for some years and have still not succeeded in doing so, except, the things he wants to change himself. But it is also difficult to change one's own behaviour. Just think of all the good intentions that fall by the wayside soon after the New Year; changing habits is difficult. A crisis can help, or as Winston Churchill said: 'Never waste a good crisis'. And mobility is no different. If a strike is announced, if there are large construction works, you see time and again that people find other ways of getting around and the traffic evaporates, as it were[175]. The coronavirus also appears to be able to magically change the way we travel and do tourism. It has also taught us to avoid plenty of travel by resorting to video-calling, teleworking, organising an e-apero, taking online courses, etc. The question is how long these behavioural changes will last.

It is important to know that changing behaviour is not just a question of knowledge. Although, it is very important, it is not sufficient. We need to act on a deeper, emotional level and make the path towards it easier.

Heath and Heath[176] offer a nice metaphor to illustrate this point: if you want to change an elephant's path, you have to talk to the rider (our brain) but also to the elephant (our emotions) and make the new path much easier and more attractive than the old one. After all, the elephant is full of fears: Am I going to be able to do that on public transport? And how am I going to recharge such an electric car? Besides, if the new behaviour is much more complicated than the previous one, you won't be able to keep it up for long.

 The metaphor is clear: we may know that it is better to recycle or to cycle to work, but in order to change our behaviour and sustain the change, it has to be emotionally rewarding and easy to follow. If recycling is very complicated or cycling is very dangerous, the chances of pursuing these goals are significantly lower.

While the literature has made a significant contribution to understanding inaction on climate change[177] little is known about the ways in which we can stimulate action through different institutional settings and interventions.

Let us therefore take a closer look at the three aspects of the elephant metaphor:

The rider: knowledge

Awareness of climate change is high. The 2021 Eurobarometer climate study reported that 93% feel climate change is a serious problem and about 58% believe businesses and industries should be responsible for tackling climate change[178]. This awareness is increasingly being felt and translates into a wider range of sustainable products and services[179]. On the other hand, the climate is becoming more and more polarised, and many people are pulling out. The actual knowledge about the theme and what to do about it is very low. So, a book like this can help ;-)! Also, education at all possible levels can help spread the knowledge further. For example, I organise annual bootcamps at the university (short but very intensive courses where all climate themes are discussed). But as I said, in addition to this knowledge, it has to sink in further.

The elephant: emotions

The elephant will only change permanently if all the emotions associated with the change are also considered. Dorit Kerret from the psychology department at Tel Aviv University works on behaviour change for sustainability from the perspective of positive psychology. What she notices around emotions is that it is important that the person has a sense of hope and self-control, which leads to commitment on an individual level. Often, with the climate crisis, people get the idea that there is no hope, and that it doesn't matter what they do anyway. The second element includes the conviction that one can successfully achieve the chosen goal. You need to believe that you can do it, as it were, or else you won't start. For example, many people are afraid to go to the various appointments of the day without a car because they are afraid of not being able to find their way by bicycle or not trusting public transport.

The path

Subsequently, the new way of moving around should actually be more logical than when you were still moving around in an unsustainable way. Positive initial experiences are very decisive in determining whether a change in behaviour will be sustained[180]. For example, this summer some Thalys trains were blocked, leaving people in the train with no water and no fresh air[181]. Strikes, too, can hamper the use of public transport. These, of course, stop people from fully embracing public transport. The government and transport organisers have an important role to play here in making the sustainable path indeed the logical option. If the buses are also stuck in traffic because there is no bus lane, why would people switch? If the price of an airplane trip is so much cheaper than the train, you have to be very committed to go by train. If cycling is like a kamikaze trip, why would you risk your life? So, the sustainable ways have to be developed a lot, so that the new path is a stayer for many people.

Dorit Kerret's research also shows that the trust that other members of society follow the same desired goal as you is important for behavioural change.

It is indeed important to realise that people can be addressed on an individual level, but that they are embedded in a socio-cultural environment, with certain values and norms about sustainability. They will also move in a physical environment, which may or may not invite sustainable behaviour. In an urban environment with well-developed public transport and many subsystems, it is certainly easier to leave the car on the sidewalk. Finally, there is also the policy environment, the regulation. We have already discussed this in the chapter Acceleration, and it should therefore also create the right framework and stimulate sustainable behaviour in a logical way. It should be noted that these policies should also take into account what we know about behaviour. For example, people do not like to lose. They will value a 10-minute increase in the duration of a journey much more highly than a service that is 10 minutes faster. Changing an acquired right, or 'taking away' what you are used to, is much harder to digest than getting something extra. That is why it is important to look carefully at the possible solutions and at what people get rather than emphasising what they will lose. With road pricing, for example, it is important to point out the benefits, such as shorter journey times and better public transport, rather than focusing on the higher cost of driving for some.

One of the most striking results in economics is that people can even be negatively influenced by a monetary reward. Think, for example, of the sorting of waste. The moment you get paid for it; the intrinsic motivation seems to fall away. Therefore, working with price instruments only is not always advisable.

FACTOR 8

Based on all these elements, I developed the 6 E's model:

$$\text{Impact} = (E{*}E{*}E{*}E{*}E) \ ^{\text{Embrace}}$$

Impact on a sustainable lifestyle =
Estimate × Engage × Empower × Enable × Encourage
Elements of the positive sustainable change programme[182].

The programme starts with setting targets *(estimate).* In order to evaluate the extent of the specific impact of an activity, an institution, a company, or a sector on the physical environment and set compatible targets, a comprehensive estimate must be made. For instance, indicating what the CO_2 footprint of the company is and then indicating by when you want to reduce the CO_2 impact. Secondly, you are going to involve everyone (emotionally) in achieving the desired goal *(engage).* The third part of the method is to *empower* with the possible options for achieving the desired goal. For example, sufficient information must be provided about shared systems, public transport formulas or Mobility as a Service subscription. Next, making it possible to achieve the goal is also part of the plan. This means that the way to reach the goal must be made easy (enable). Finally, the participants must be encouraged to pursue the set goal by rewarding them in a positive way *(encourage).*

It has been put into an equation to show that if you forget one of the basic 5Es, the impact will be zero.

A 6th E, *embrace,* can be added. Embracing the differences between people. So, when you set up a project, take a good look at whether it is sufficiently geared to the entire target group or whether it is best to divide it up between groups[183].

The model can, for example, be used to set up campaigns. This year I set up the campaign '30 days with less cars' in the month of June. It was a campaign rolled out in Flanders together with the Network Sustainable Mobility.

In order to *engage* people, we gave them four good reasons to join the campaign: for your health, for the adventure, for your wallet, and for the change. These reasons play on different type of emotions that could let them choose to be willing to try it out for one month. We also made it easier: first of all, by empowering them by giving information and tips on the different alternatives; and by enabling it through the sponsorship of different mobility players like the public transport operators and sharing services that made it possible for those who joined the action, to have free ridership during the month of June. A survey helped estimate the impact: 25% of the people that subscribed for the challenge did not use the car for the complete month. Instead, they used more public transport and tried - often for the first time - a shared car. Finally, for those who successfully made it till the end, a reward in the sense of extra benefits on alternative transport modes were given.

ALL IN LOVE!

In this book, I want to show that a different, more climate-neutral mobility system is achievable. It is possible! And yes, that means a different way of getting around, of living. That change is not easy. In life, there are a few moments that make it possible to change; where clinging to your traditional behaviour is no longer tenable and where you automatically start to transform.

Firstly, there are the major life crises, for example when you lose your job, suffer a burnout or leave your partner. At these moments everything is called into question and that is what makes it possible to approach things in a different way. The big pause button that was pressed during the lockdown of the corona crisis was also such a significant moment. It gave many people new insights about how they live, how they move, and what is essential. It just gave that shock effect, and we were forced to experience life in a different way. There was a lot of beauty in that lockdown. Some of our behavioural changes should be continued to move towards a climate-neutral mobility system. We learned to stay in our own neighbourhood, close to our own homes. We have learned to walk and shop in the vicinity. We have learned to telecommute and to make video calls. Even the not so obvious hobbies, such as dancing in my case, became online activities. What beautiful lessons in avoidance! Yes, so much can be done without having to move. We learned to travel within our own country. And yes, we are learning that there is so much beauty to see close by. And ultimately, it is also a lesson in not always looking externally but knowing that there is so much to experience within us and in our own home.

The lockdown has also caused people to jump on their bikes en masse. And as a result, new infrastructure was swiftly provided for cyclists. Walking and the public space required for it also became a clear priority. And even public transport, which will not be prioritised for a while, has had the advantage of much better performance because the buses and trams were not stuck between cars. There are also opportunities in the corona crisis for change makers. The support that certain businesses asked for can be maximally focused on those companies that can help realise the climate-neutral society of the future. Thanks to Corona, what we have experienced has taught us that it is possible to live in a city with

70% less air pollution, less noise pollution from both cars and planes and that a change in behaviour is possible. This gives so much hope that the transition necessary for the climate crisis is also possible. Meanwhile, our policy makers have understood that taking drastic measures is sometimes necessary to save lives. So yes, change can come from big crises.

Secondly, being in love turns out to be a powerful lever for change. Try to reimagine that feeling of falling head over heels for someone, or maybe you are in love right now!? For some, that feeling is as bad as a major life crisis, but for many it is also a moment where they jump around, do crazy things, change their lifestyle because they think the other one will like it. That's the right attitude to put my words into action. Just try it for a week. Imagine sitting on the bus with a feeling of love. A bit exciting but also a lot of fun. Or jump on your bike. Enjoy that wonderful feeling of freedom, of getting somewhere with the wind in your hair and under your own steam; in love with life, with the world and with living in a different way. I am sure it will work out!

Epilogue

I learned that courage is not the absence of fear, but the overcoming of it. The brave man is not he who is not afraid, but he who overcomes that fear.

Nelson Mandela

Why is the fear of the corona virus so different from the fear of climate change? Or, in other words, why does one thing lead to drastic measures and the other doesn't? Certainly, it has to do with the time horizon.

With corona, people are dying from the virus now, demonstrably. With climate change, all kinds of phenomena like hurricane, extreme heat, drought, famine and what not, will eventually cause people to die. But it is less direct and less rapid, even so, equally real demonstrable and proven.

Just like with the climate change, with Corona, you can see that some people feel addressed, and others do not. It is a human thing; "if I am not affected, if I do not feel it *tout court,* then I am not going to do anything about it". But the fear of the coronavirus is there, and people are afraid, for themselves and for their loved ones.

Suddenly, there is that generation gap again; this time it was the other way round. The younger generation did not feel so addressed by corona: "I am strong enough, I will survive". So, the older generation started calling them to order and telling them to be responsible. Does that ring a bell? Isn't that what is happening with climate change? The younger generation calling the older generation to responsibility? And does the older generation not really feel involved in climate change, because ' the old model will last my time'?

The fact is that measures are being taken for corona. Drastic measures! Unseen measures! What about measures for the climate? There we heard: "It will cost too much to our economy", "People still need to be able to move around, don't they?" and "the technology will fix it".

Experts warned that it would cost a lot more if nothing was done, that there were quite a few interesting options to explore. But for some reason, they were not listened to; there had to be support first. So that support turns out to be fear? If everyone is afraid, apparently there is the possibility of introducing things. Amidst the Corona crisis, it seemed the government is now taking the experts' advice and implementing it. But the many voices and all the advice that have been given over the years were ignored, which is why we are now facing such a mess. Does the climate crisis have to be like that first: *just that little bit too small and too late?*

The consequences of the lockdown for the economy were catastrophic. Many people and companies are still facing serious problems. Do we individually or collectively have the buffer to cope with this? I would like to believe so. Just as we have the buffer to make the transformation for climate change.

I have always seen fear as something to be avoided, something that you should not feel and if you feel it, there is something wrong with you. But fear, like all emotions, is there to signal you something. And then you see how beautiful fear is; it is about love. Fear indicates that you are afraid to lose something that you love, just as anger is a reaction to protect what you love.

If we look at it this way, then perhaps the explanation lies in why the fear of the corona crisis is different from that of the climate crisis. The corona crisis touches directly on what we like to see: *me, myself and I,* and oh yes, my old mother. The climate crisis, however, does less of that. Could it be that we have become so individualistic and alienated from nature that our love for it and the others affected by the climate crisis is so much smaller? The idea of "it won't be me, it will be the people in Africa who will suffer the most. I'll put on my air conditioning in a hot summer and my central heating in a cold windy winter. And my love for that wider field is just a little less now".

Can we transcend this primary fear and try to act from love right now? Or do we wait until we have to react again in a hurry, and are dominated by fear?

A final remark: when would you leave a bad situation? For example, at work, or in your love life. Would that be when it is 90% bad or 20%? Many people stay, even for their whole life, in a situation that is not 100% good. They wait, and wait, and are suffering for years because of the fear of change. But at a certain moment, when their drinking partner, also cheat on them, then they might find the courage to leave. At work, it is the same. How many burnouts should there be, to see something is not going well? Are we waiting for the climate crisis to be so bad before we finally act? Or can we take action right away as we can already see the bad situation we are getting in?

Brussels, August 2022

Acknowledgements

This book was a turning point in my life. After several years of full participation in all possible climate expert committees, my story about the 8As and Factor 8 took further shape. In the years before, I had mainly dealt with the first 4As. But it is clear that a government policy is also needed to speed things up, to connect actors and to create the right framework to motivate us all to change.

It was a turning point because I was very much aware of the urgency and so many initiatives came about or were started by me to indicate the urgency of the climate crisis. And when everything becomes too much, the only thing you can do is stand still. The COVID crisis also pushed the stop button. That was very frustrating, but it also gave me the opportunity to take more time for this book, to reflect on the message I wanted to convey and the way I wanted to do that. I am extremely grateful that the publishing team, and in particular Karel Puype and Katrien Van Moerbeke, gave me all that space and encouraged me.

The collaboration with Mathilde Guegan was also very nice. It really became a co-creation in which we found each other in the visualisation of the message that the book wants to bring. Ganiyat Temidayo Saliu helped with the translation. It was really great to work with her!

Also, many thanks to my research group Mobilise. It is, after all, a story of all the knowledge we gather there together. The support of all of them and their research is the basis of this book. For the English translation and update I especially want to thank Imre Keserü and Koen Mommens. As the team leaders on urban mobility and sustainable logistics they provided me with the latest updates from their projects. Also, many thanks to the other experts of the climate commissions. Not only for their valuable feedback but also for their incredible commitment.

I would also like to thank the Vrije Universiteit Brussel. At the VUB, we are practicing what we preach and making the move to shape our activities in a more sustainable way. Experiencing this transition and seeing the obstacles and levers to change, as well as the enthusiasm of many people, gives this story additional practical insights.

I would also like to thank the people of the Brussels Mobility Commission, which I have chaired for some time now. Through the debates we have there and the contacts with the minister and the cabinet we stay in touch with the field and the realities of everyday life, because knowledge is not gained in an ivory tower. The MOW department of the Flemish government and the MORA council are also a constant source of fine exchanges.

I would also like to thank my children, Ilya and Hanne. As twenty-some-things, they teach me a lot about how they look at the climate issue, but also how they want to and will move into the future. Our discussions at the kitchen table teach me a lot and I hope especially for them that the tide will turn, and they will be able to live in a world that respects the boundaries of the ecosystem. But I also had my mother in mind when I wrote this book. In the end, we want all generations to come along, and a transformation must be possible for the not-so-young generation as well.

ACKNOWLEGMENTS

FACTOR 8

Endnotes

[1] NASA – Earth Observatory. https://earthobservatory.nasa.gov/world-of-change/
global-temperatures#:~:text=According%20to%20an%20ongoing%20
temperature,1.9%C2%B0%20Fahrenheit)%20since%201880.

[2] IPCC, AR5, chapter 12. The climate action tracker is used to monitor current policy
ambitions. This leads to a scenario of 3 to 4 degrees (Climate Action Thermometer).

[3] World Meterological Organisation 2022
https://public.wmo.int/en/media/press-release/wmo-update-5050-chance-of-global-
temperature-temporarily-reaching-15%C2%B0c-threshold

[4] For Belgium, the largest contribution to total emissions comes from CO_2, accounting
for 85.1% in 2016. Emissions of CH4 (methane) make up the second largest share with
6.8%, followed by emissions of N2O (nitrous oxide) with 5.1%.

[5] Mommers, Jelmers, How are we going to explain this. Our future on an increasingly
warm planet, De Correspondent, 2019

[6] CO_2 equivalents are usually used, with CO_2 as the reference gas against which other
greenhouse gases are measured.

[7] https://gml.noaa.gov/webdata/ccgg/trends/co2/co2_annmean_mlo.txt

[8] EU Crop Market Observatory (DG AGRI and JRC Mars Bulletin (2018)

[9] Wim Thiery, Climate Bootcamp, VUB, 2019

[10] Robert M. DeConto et al. "Contribution of Antartica to past and future sea-level rise,
Nature, vol. 531 (2016), PP. 591-597.

[11] Global Carbon Budget 2018

[12] Global Warming of 1.5°C. An IPCC Special Report on the impacts of global warming
of 1.5°C above pre-industrial levels and related global greenhouse emission pathways,
in the context of strengthening the global response to the threat of climate change,
sustainable development, and efforts to eradicate poverty, IPCC (2018). chapter 2,
p. 158.

[13] EC, 2014

[14] European Commission
https://ec.europa.eu/clima/eu-action/european-green-deal/2030-climate-target-
plan_en#:~:text=With%20the%202030%20Climate%20Target,below%201990%20
levels%20by%202030

[15] European Commission – European Green Deal
https://ec.europa.eu/clima/eu-action/european-green-deal_en

[16] EEA, 2017; ICCT, 2018 This includes both direct and indirect CO_2 emissions and
combines both ETS & non-ETS sectors in the EU for 2015. Other sources often
calculate 25%. This is then without domestic flights and only direct CO_2 emissions.
Direct emissions are based on the geographical boundaries but do not take into
account the emissions emitted elsewhere in the world by the consumption of the
inhabitants.

[17] De Standaard, 2019. *Billions For Sustainable International Shipping*
https://www.standaard.be/cnt/dmf20191218_04772862

[18] European Commission, https://ec.europa.eu/clima/eu-action/transport-emissions/

reducing-emissions-aviation_en#:~:text=The%20aviation%20sector%20creates%20
13.9,GHG%20emissions%20after%20road%20transport.

19 International Civil Aviation Organisation

20 Kharina, A., & Rutherford, D. (2015). Fuel efficiency trends for new commercial jet
    aircraft: 1960 to 2014. https://theicct.org/sites/default/files/publications/ICCT_
    Aircraft-FE-Trends_20150902.pdf

21 The World Bank Data. *Air transport, Passenger carried.* https://data.worldbank.org/
    indicator/is.air.psgr

22 https://data.worldbank.org/

23 Brussels Airlines pays around 10 million euros annually

24 EG, 2016; EER 2016

25 In Belgium, industrial emissions were reduced by improvements in energy efficiency
    and by closing large industrial plants in the period 1990-2016, especially in the
    Walloon Region. In the energy sector, a fuel switch from coal to gas resulted in lower
    emissions, and in the waste sector the reduction of methane proved important (A net-
    zero Greenhouse Gas Emissions. Belgium, 2050. Initiating the debate on Transition
    policies): a report written by Pieter Boussemare, Jan Cools, Michel de Paepe, Erik
    Mathijs, Bart Muys, Karel Van Acker, Han Vandevyvere, Arne van Stiphout, Frank
    Venmans, Kris Verheyen, Pascal Vermeulen, Sara Vicca and myself under the direction
    of Tomas Wyne.

26 European Road Federation. European Road Statistics 2021. https://erf.be/statistics/
    general-data-2021/

27 https://www.uber.com/newsroom/uberkittens/

28 https://www.ipc.be/services/markets-and-regulations/e-commerce-market-insights/e-
    commerce-articles/parcel-lockers-2022

29 Mommens, K., Buldeo Rai, H., Macharis, C., 2021, "Delivery to homes or collection
    points? An environmental impact analysis for urban, urbanised and rural areas in
    Belgium", Journal of Transport Geography, Vol. 94, 103095

30 Survey of travel behaviour 5.3, 2018 and FPS Mobility & Transport, 2019. There is quite
    some difference between the 3 regions. For example, people in Wallonia travel more by
    car than in Flanders, but in Brussels they travel much less. Depending on the activity,
    people also use the car more or less. For work and shopping, the car is used most.
    For leisure, most people carpool.

31 Federal Planning Bureau, 2019 Expressed in tonne-kilometres which is the
    multiplication of distance driven and mass transported.

32 EU Transport In Figures. Statistical Pocketbook 2021. https://op.europa.eu/en/
    publication-detail/-/publication/14d7e768-1b50-11ec-b4fe-01aa75ed71a1

33 ACEA 2022. *New Car Registrations By Fuel Type, European Union.* https://www.
    acea.auto/files/20220720_PRPC-fuel_Q2-2022_FINAL.pdf

34 https://fresh-trending.com/15-european-international-locations-have-15-electrical-
    automotive-gross-sales-new-automotive-gross-sales/

35 Statbel, 2019. ACEA

36 Eurostat Statistics Explained.
    https://ec.europa.eu/eurostat/statistics-explained/index.php?title=Passenger_
    cars_in_the_EU#An_almost_10_.25_increase_in_EU-registered_passenger_cars_
    since_2015

37 https://www.best-selling-cars.com/europe/2021-full-year-europe-new-car-sales-and-
    market-analysis/

38 https://www.euronews.com/2018/02/07/which-european-commuters-spend-the-most-
    time-in-traffic-jams-

39 Vlaams Verkeerscentrum, 2019. Traffic indicators: http://indicatoren.verkeerscentrum.
    be/vc.indicators.web.gui/indicator/index#/presentation-tab-table last visited
    19/12/2019

40 European Commission, Impact Assessment for the White Paper "Roadmap to a Single
    European Transport Area - Towards a competitive and resource efficient transport
    system" SEC(2011) 358 final, https://eur-lex.europa.eu/LexUriServ/LexUriServ. do?uri
    =SEC:2011:0358:FIN:EN:PDF

41 Liu, N. Miyashita, L., Mcphail, G., Thangaratinam, S., Griff, J., 2018. Late Breaking
    Abstract - Do inhaled carbonaceous particles translocate from the lung to the
    placenta? European Respiratory Journal, 52, PA360.

42 EEA, 2019

43 WHO, 2014; EEA, 2016; ECA, 2018

44 https://openknowledge.worldbank.org/bitstream/
    handle/10986/36501/9781464818165.pdf?sequence=4&isAllowed=y

45 https://www.eea.europa.eu/highlights/industrial-air-pollution-in-europe

46 VMM, 2017

47 European Mobility Atlas 2021. https://eu.boell.org/sites/default/files/2021-02/
    EUMobilityatlas2021_FINAL_WEB.pdf

48 https://ec.europa.eu/eurostat/databrowser/view/tran_sf_roadve/default/
    table?lang=en

49 https://www.eea.europa.eu/ims/greenhouse-gas-emissions-from-transport

50 COP 21, Paris

51 https://ec.europa.eu/energy/en/data-analysis/energy-modelling/eu-reference-
    scenario-2016

52 European Environment Agency 2019

53 https://ourworldindata.org/co2-emissionshttps://ourworldindata.org/co2-emissions

54 https://www.statistiekvlaanderen.be/nl/koolstofvoetafdruk

55 https://germany.myclimate.org/en/flight_calculators/new

56 Buldeo Rai, Mommens, Verlinde & Macharis (2019). https://www.mdpi.com/2071-
    1050/11/9/2534

57 Buldeo Rai, Verlinde & Macharis (2018) The "next day, free delivery" myth unravelled:
    Possibilities for sustainable last mile transport in an omnichannel environment.
    International Journal of Retail & Distribution Management, 47(1), pp.39-54.

58 Doctorate of Heleen Buldeo Rai, "The Environmental Sustainability of the Last Mile in
    Omnichannel Retail", VUB-MOBI.

59 Messagie, M., Boureima, F.-S., Coosemans, T., Macharis, C. en J. Van Mierlo, 2014,
    "A Range-Based Vehicle Life Cycle Assessment Incorporating Variability in the
    Environmental Assessment of Different Vehicle Technologies and Fuels", Energies, 7,
    pp. 1467-1482, DOI: 10.3390/en7031467 (2.072: JCR 2014).

60 Van Mierlo, J., Messagie, M., & Rangaraju, S. (2017). Comparative environmental assessment of alternative fueled vehicles using a life cycle assessment. *Transportation research procedia*, 25, 3435-3445.

61 figures from the FPS Public Health, Food Chain Safety and Environment and the National Energy and Climate Plan (p.90). Mandatory blending percentages are provided for in this plan of 8.4% from 2020 until 2024, after which the aim is to increase blending to 14% in 2030. https://ec.europa.eu/research-and-innovation/en/horizon-magazine/why-raising-alcohol-content-europes-fuels-could-reduce-carbon-emissions#:~:text=Most petrol now sold at,%25 bioethanol, known as E10

62 In the new European Directive (RED II), biofuels may still be counted as renewable energy. Also, first generation biofuels.

63 European Commission, Directorate-General for Mobility and Transport, Essen, H., Fiorello, D., El Beyrouty, K., et al., *Handbook on the external costs of transport: version 2019 – 1.1,* Publications Office, 2020, https://data.europa.eu/doi/10.2832/51388, https://op.europa.eu/en/publication-detail/-/publication/9781f65f-8448-11ea-bf12-01aa75ed71a1

64 van Lier, T., C. Macharis and J. Van Mierlo, 2017. External Costs of Transport, in C. Macharis and J. Van Mierlo (eds), *Sustainable Mobility and Logistics* (85-121). VUBPress, Brussels.

65 Koen Mommens, Nicolas Brusselaers, Tom van Lier, Cathy Macharis, A dynamic approach to measure the impact of freight transport on air quality in cities, Journal of Cleaner Production, Volume 240, 2019

66 Committee of the Carbon Pricing Leadership Coalition (CPLC)

67 van Essen, H., van Wijngaarden, L., Schroten, A., Sutter, D., Bieler, C., Maffii, S., ... El Beyrouty, K. (2019). *Handbook on the external costs of transport: Version 2019.* Delft: CE Delft. https://doi.org/10.2832/27212

68 In more recent studies, these costs are estimated to be higher than in earlier studies. This is mainly due to an increasing knowledge of the subject and sensitivity risks, which in turn translates into more sophisticated modelling. However, a large spread is still noticeable in the estimation of damage costs, which indicates the uncertainty that these approaches entail. Kuik et al. (2009), in a study strongly based on the United Nations Framework Convention on Climate Change (UNFCCC), propose a range of 69-241 €/t CO_2 equivalent to a central value of 129 €/t CO_2 equivalent for the year 2025 and 128-396 €/t CO_2 equivalent with a central value of 225 €/t for 2050 (base year 2005). The Ricardo-AEA study (2014) provides an updated handbook of Maibach et al. (2008) with a central value for the carbon price of € 90/t CO_2 equivalent (€ 48-168, 2010 prices), which is comparable to other studies. For example, UBA (2012) recommends a central value of € 80/t CO_2 equivalent (with a range of € 40-120) as a guideline for Germany, and Watkiss and Downing (2008) recommend £ 80/t CO_2 equivalent for the UK (for 2010).

69 https://carbonpricingdashboard.worldbank.org/map_data

70 https://www.standaard.be/cnt/mf20190201_04147117?articlehash=B8F7DE9FB7A5C310886815F786B3F8B968360E21894AC4142005E44AC358D8C4935FB64D4376A0EB7B465638EAB086384FB3AAC5CFF944E59751676BACC09903

71 Schroten, A., Van Essen, H., Ban Wijngaarden, L., Sutter, D., & Andrew, E. (2019). *Sustainable Transport Infrastructure Charging and Internalisation of Transport Externalities: Executive Summary.* https://doi.org/10.2832/246834

ENDNOTES

FACTOR 8

72 Studio 012 Secchi/Vigano in Brussels Capital Region (2014)

73 Study by VITO, Common Ground and VRP for the Flemish government: Monetising the impact of urban sprawl in Flanders, 2019

74 Research Movement Behaviour Flanders 5.3

75 Acerta. (2016) Sustainable mobility formulas are on the rise. Press release. Retrieved from http://consult.acerta.be/nl/pers/duurzame-mobiliteitsformules-zitten-de-lift

76 The same study also calculates the costs of infrastructure (maintenance of roads, utilities (water, gas, electricity, sewage) and lighting), the loss of open space and ecosystem services and public services. All in all, by choosing a scenario in which open space is returned, 25.6 billion euros can be saved by 2050. A scenario in which the take-up of open space is halted could save €15.9 billion.

77 https://vito.be/nl/nieuws/maatschappelijke-kosten-van-urban-sprawl-voor-het-eerst-becijferd

78 Van Lier, T., De Witte A. and C. Macharis, 2014, 'How worthwhile is teleworking from a sustainable mobility perspective: the case of Brussels Capital region', *European Journal of Transport and Infrastructure Research (EJTIR), 14(3),* pp. 240-263 (0.818: JCR 2014).

79 Anderson, A. J., Kaplan, S. A., & Vega, R. P. (2015). The impact of telework on emotional experience: When, and for whom, does telework improve daily affective well-being? *European Journal of Work and Organizational Psychology, 24(6),* 882-897. https://doi.org/10.1080/1359432X.2014.966086

80 Indicators of Energy Use and Efficiency. International Energy Agency. Paris, France (1997)

81 Car occupancy is an indicator of the number of people travelling by car. The average occupancy rate in Europe ranges from 1.1 to 1.2 for commuting to work and 1.4 to 1.7 for family travel (European Environment Agency, 2016), while the capacity of passenger cars is typically 4 to 5 people.

82 Commuting diagnostics of the FPS Mobility and Transport

83 Rayle, L., Dai, D., Chan, N., Cervero, R., & Shaheen, S. (2016). Just a better taxi? A survey-based comparison of taxis, transit, and ridesourcing services in San Francisco. *Transport Policy, 45,* 168-178. https://doi.org/10.1016/j.tranpol.2015.10.004

84 Hampshire, R.C., Simek, C., Fabusuyi, T., Di, X., & Chen, X., 2017, *Measuring the Impact of an Unanticipated Suspension of Ride-Sourcing in Austin, Texas* (SSRN Scholarly Paper No. ID 2977969). Social Science Research Network. Available from https://papers. ssrn.com/abstract=2977969

85 Schaller, B., 2017, Unsustainable? The Growth of App-Based Ride Services and Traffic, *Travel and the Future of New York City.* Available at: http://www.schallerconsult.com/ rideservices/unsustainable.pdf

86 Buldeo Rai, H., 2021, "Duurzaam online shoppen – praktijkgids voor e-commerce van morgen", Lannoo

87 In the long term, a physical Internet is envisaged in which all information is shared and in which it does not matter who transports the goods, as long as it reaches its final destination in the most sustainable way. For more information see: https://www.etp-logistics.eu

88 Mansuy, J. (2022). Circular value creation: An incumbent firms' perspective on the circular economy transition

[89] Colville-Andersen (2014)

[90] http://www.leuvenautovolenautovrij.be

[91] Jones, P. et al. , CREATE project http://www.create-mobility.eu

[92] http://www.copenhagenize.com/2014/09/the-arrogance-of-space-paris-calgary.html

[93] Keseru, I., Wuytens, N., de Geus, B., Macharis, C., Hubert, M., Ermans, T., & Brandeleer, C. (2016). *Monitoring the impact of pedestrianisation schemes on mobility and sustainability* (pp. 97- 106). BSI - Brussels Centre Observatory | BSI-BCO. Retrieved from http://bco.bsi- brussels.be/monitoring-the-impact-of-pedestrianisation-schemes-on-mobility-and-sustainability/

[94] Wallström, M. (2004). Reclaiming cuty streets for people. Chaos or quality of life? European Commission. Retrieved from http://ec.europa.eu/environment/pubs/pdf/streets_people.pdf

[95] PMU, 2014

[96] B. de Geus, N. Wuytens, T. Deliens, I. Keserü, C. Macharis, R. Meeusen, Psychosocial and environmental correlates of cycling for transportation in Brussels, Transportation Research Part A: Policy and Practice, Volume 123, 2019.

[97] https://www.bike-eu.com/36156/market-report-belgium-e-bikes-keep-growing

[98] The STOP principle gives policy-makers and planners a guide to prioritising the most environmentally friendly ways of getting around: *first Stappers, then Trappers, then Public and finally Private Transport*. And it is this priority that is to be followed in planning and corresponding investments.

[99] However, the principle of P&R is also much criticised because it often generates more car kilometres than it saves and it also takes up space at the expense of densification around public transport hubs.

[100] https://parkride.eu/

[101] Bates, J., & Leibling, D., 2012, *Spaced Out Perspectives on parking policy.* RAC Foundation. Available from http://www.racfoundation.org/assets/rac_foundation/content/downloadables/spaced_out-bates_leibling-jul12.pdf [last accessed 4/03/2020].

[102] Results of studies vary considerably. They range from 2.5 (Douma and Gaug 2009) to 13 (Martin et al. 2010). *(8) (PDF) Carsharing: the impact of system characteristics on its potential to replace private car trips and reduce car ownership.* Available from: https://www.researchgate.net/publication/328227188_Carsharing_the_impact_of_system_characteristics_on_its_potential_to_replace_private_car_trips_and_reduce_car_ownership [accessed Jan 20 2020].

[103] A third type of shared car is the one that is lent out among neighbours (peer-to-peer).

[104] Wiegmann, M., Keseru, I. and C. Macharis, 2019, "The development of carsharing in the Brussels Region between 2016 and 2018", MOBI-report.

[105] All information on car sharing and ride sharing for Belgium can be found at: https://www.taxistop.be

[106] (Brussels Mobility, 2019)

[107] https://www.thebulletin.be/road-safety-institute-concerned-number-e-scooter-accidents

[108] Ramboll, Whimimpact: Insights from the world's first Mobility-as-a-Service (MaaS) system, 2019.

[109] Storme, T., De Vos, J., De Paepe, L., & Witlox, F. (2020). Limitations to the car-substitution effect of MaaS. Findings from a Belgian pilot study. *Transportation Research Part A: Policy and Practice, 131,* 196-205.

[110] Special Eurobarometer 495. Report. Mobility and Transport. European Union, 2020

[111] Mommens, K., Lebeau, P., Verlinde, S., van Lier, T., Macharis, C., 2017. Evaluating the impact of off-hour deliveries: an application of the Transport Agent-Based Model. Transportation Research Part D Transport and Environment, Vol. 62, pp, 102-111.

[112] Mommens, K. M. (2019). The development of an assessment framework for multimodal freight transport of different cargo types in Belgium.

[113] Mommens, Koen (2019). The development of an assessment framework for multimodal freight transport of different cargo types in Belgium. phd VUB-MOBI.

[114] "Synchromodal Transport and the Physical Internet", PhD Tomas Ambra, VUB-UHasselt

[115] https://ce-center.vlaanderen-circulair.be/nl/publicaties/publicatie-2/6-impact-of-circular-economy-on-achieving-the-climate-targets-case-mobility

[116] Quentin De Clerck, Tom van Lier, Maarten Messagie, Cathy Macharis, Joeri Van Mierlo, Lieselot Vanhaverbeke, Total Cost for Society: A persona-based analysis of electric and conventional vehicles, Transportation Research Part D: Transport and Environment, Volume 64, 2018, Pages 90-110.

[117] https://www.cnbc.com/2022/05/18/ev-battery-costs-set-to-spike-as-raw-material-shortages-drags-on.html https://www.cnbc.com/2022/05/18/ev-battery-costs-set-to-spike-as-raw-material-shortages-drags-on.html

[118] Bloomberg NEF (2021)

[119] This is partly due to stricter regulations introduced by the European Commission, which stipulate that each car sold by a car manufacturer must emit on average only 95 g CO_2 /km. At the moment, that maximum is 120 g/km. The result is that car manufacturers, if they still want to sell their more polluting cars, have every interest in also selling a lot of electric cars in order to reach an average emission of 95 g CO_2 eq.

[120] Cozzi, L., & Petropoulos, A. (2021). Global SUV sales set another record in 2021, setting back efforts to reduce emissions. https://www.iea.org/commentaries/global-suv-sales-set-another-record-in-2021-setting-back-efforts-to-reduce-emissions.

[121] You can do that either in a normal socket but that is quite slow (10h) but you can also install a charging station in your garage and then the charging time is about 4h. Some are equipped with smart meters, which allow you to send the bill for charging to your employer. More info on: Powerdale.com

[122] An overview of public charging stations can be found here: https://nl.chargemap.com/map. To use such a charge point, you need a pass.

[123] https://www.sparkcharge.io/

[124] https://www.creg.be/sites/default/files/assets/Publications/Studies/F1609EN.pdf

[125] Powerdale.com

[126] This piece is largely based on the text I made with my colleague Lieselot Vanhaverbeke for the book *Homo roboticus: 30 questions and answers about man, robot & artificial intelligence,* VUBPress, Brussels, ISBN 9789057188503, p. 296.

[127] Fagnant, D. & K. Kockelman, 2015, Preparing a nation for autonomous vehicles: opportunities, barriers and policy recommendations, Transportation Research Part A: Policy and Practice, Volume 77, Pages 167-181. Litman, T. (2017). *Autonomous vehicle implementation predictions* (p. 28). Victoria, BC, Canada: Victoria Transport Policy Institute.

[128] Firnkorn & Müller, 2015 Firnkorn, Joerg & Müller, Martin. (2015). Free-floating electric carsharing-fleets in smart cities: The dawning of a post-private car era in urban environments? Environmental Science & Policy. 45. 10.1016/j.envsci.2014.09.005.

129 https://mobi.research.vub.be/en/first-ride-of-a-self-driving-shuttle-bus-on-brussels-health-campus

130 STIB is also running tests in Woluwe, Vias in Han-sur-Lesse and Waterloo, and De Lijn is going to test at Zaventem airport.

131 Harb et al, 2018

132 Wadud, Z., MacKenzie, D. & P. Leiby (2016), Help or hindrance? The travel, energy and carbon impacts of highly automated vehicles, Transportation Research Part A: Policy and Practice, Volume 86, 2016,Pages 1-18. and Fagnant, D. & K. Kockelman, 2015, Preparing a nation for autonomous vehicles: opportunities, barriers and policy recommendations, Transportation Research Part A: Policy and Practice, Volume 77, Pages 167-181.

133 *Jingjun Li et al, 2019,* A Systematic Review of Agent-based Simulations for Assessing the Impact of Vehicle Automation within Mobility Networks, VUB-MOBI

134 Viegas, J., & Martinez, L. (2017). Transition to Shared Mobility (pp. 1-55). International Transport Forum.

135 COWI and PTV, The Oslo Study-How autonomous cars may change transport in cities, April 2019

136 Jingjun Li, Evy Rombaut, Koen Mommens, Cathy Macharis, and Lieselot Vanhaverbeke, 2020, Towards an Integrated Mobility and Logistics Network for Autonomous Vehicles: Review and Opportunities of the Agent-based Approach, MOBI paper for the HEART conference

137 https://www.sfmta.com/sites/default/files/projects/2016/SF%20Smart%20City%20Challenge_Final.pdf

138 Vias, 2017

139 The Guardian, 11/12/2019

140 EUROCONTROL Comprehensive Aviation Assessment. https://www.eurocontrol.int/publication/eurocontrol-comprehensive-aviation-assessment

141 https://www.ikea.com/us/en/newsroom/corporate-news/ikea-u-s-to-convert-its-new-york-last-mile-delivery-fleet-to-electric-vehicles-by-may-2021-pub61276adf
https://www.aboutamazon.com/news/sustainability/go-behind-the-scenes-as-amazon-develops-a-new-electric-vehicle

142 https://www.renault-trucks.com/en/electromobility

143 https://electrek.co/2018/02/08/tesla-semi-electric-semi-truck-production/

144 https://new.siemens.com/global/en/products/mobility/road-solutions/electromobility/ehighway.html

145 https://www.vdlgroep.com/nl/nieuws/ontwikkeling-elektrische-truck-met-waterstof-range-extender

146 Lebeau, P., Macharis, C. and J. Van Mierlo, 2019, "How to Improve the Total Cost of Ownership of Electric Vehicles: An Analysis of the Light Commercial Vehicle Segment", *World Electric Vehicle Journal, 10(4),* 90, https://doi.org/10.3390/wevj10040090.

147 https://www.logistiek.nl/distributie/nieuws/2017/07/autonome-robots-veranderen-complexe-last-mile-drastisch-101157077

148 Schroeder, 2018. https://www.gondola.be/nl/news/ahold-delhaize-test-zelfrijdende-minisupermarkten

149 https://www.digitaltrends.com/cars/robomarts-self-driving-grocery-store-is-like-amazon-go-on-wheels/

150 FPS Mobility, 2017

ENDNOTES

FACTOR 8

151 McKinnon, 2014

152 Study "Decarbonisation of the logistics sector in Flanders" for MOW, VUB-MOBI-VIL-SWECO, 2019.

153 https://www.theglobeandmail.com/report-on-business/rob-commentary/we-traded-carriages-for-cars-lets-embrace-the-next-disruption/article29782316/#:~:text=The%20shift%20from%20horses%20to,main%20source%20of%20city%20transport.

154 Transport and Environment (2020): Company car report, https://www.transportenvironment.org/wp-content/uploads/2021/06/2020_10_Dataforce_company_car_report.pdf

155 https://www.emta.com/IMG/pdf/barometer_2022-2.pdf?4298/bcdfdf4388d626b01345007e4ad46596fd218096, EMTA Barometer 2022 Based on 2020 Data. 16th Edition.

156 Study "Uitrol van een systeem van wegenheffing" by MOTIVITY for the Flemish Government, Department MOW.

157 For freight transport, on the other hand, there are many systems in Europe where payments are made per kilometre.

158 https://www.areacmilano.it/en#:~:text=If%20you%20live%20in%20Area,40%20free%20entrances%20a%20year.

159 https://www.mckinsey.com/industries/automotive-and-assembly/our-insights/germanys-9-euro-ticket-a-potential-solution-for-urban-mobility-issues

160 European Mobility Atlas 2021. https://eu.boell.org/sites/default/files/2021-02/EUMobilityatlas2021_FINAL_WEB.pdf

161 Macharis, C., Brusselaers, N. and K. Mommens, 2019, "Challenge for the near future: Instruments for a climate friendly use of road infrastructure", Road pricing in the Benelux: Towards an efficient and sustainable use of road infrastructure: Theory, application and policy, in: van den Berg, L. & Polak, J. (eds.), Brussels, BIVEC-GIBET jubilee book, pp. 21-44.

162 European Mobility Atlas 2021. *Facts and figures about transport and mobility in Europe.* https://eu.boell.org/sites/default/files/2021-02/EUMobilityatlas2021_FINAL_WEB.pdf

163 Luxembourg Institute of Socio-economic Research. https://liser.elsevierpure.com/en/publications/evaluation-of-an-incentive-program-to-stimulate-the-shift-from-ca

164 May, Ermans & Hooftman, 2019, see endnote 1 (endnote 1 is missing)

165 Vandenbroucke, A., Mezoued, A. M., & Vaesen, J. (2019). Commercial vehicles and sustainable mobility. Analysis and challenges. ASP Academic & Scientific Publ.

166 European Commission. https://taxation-customs.ec.europa.eu/taxation-1/excise-duties/excise-duty-energy_en

167 Grabbe, H., & Lehne, S. (2019). *Climate Politics in a Fragmented Europe.* Carnegie Endowment for International Peace. https://carnegieendowment.org/files/Lehne_Grabbe_Climate_v2.pdf

168 Indimo project, Eurostat 2020.

169 More about the MAMCA method and its software can be found at: http://www.mamca.be/en/

170 European Commission. *Best Practices In Citizen Science For Environmental Monitoring.* COMMISSION STAFF WORKING DOCUMENT. https://ec.europa.eu/environment/legal/reporting/pdf/best_practices_citizen_science_environmental_monitoring.pdf

[171] Van Brussel, Suzanne, and Huib Huyse. "Citizen Science on Speed? Realising the Triple Objective of Scientific Rigour, Policy Influence and Deep Citizen Engagement in a Large-Scale Citizen Science Project on Ambient Air Quality in Antwerp." *JOURNAL OF ENVIRONMENTAL PLANNING AND MANAGEMENT,* vol. 62, no. 3, 2019, pp. 534-51.

[172] Pappers, J., & De Wilde, L. (2022). Evaluating citizen science data: A citizen observatory to measure cyclists' waiting times. *Transportation Research Interdisciplinary Perspectives, 14,* 100624.

[173] https://theshift.be/nl/inspiratie/mobility-is-a-serious-game

[174] There are many non-profit organisations and consultancy firms that have specialised in this area, such as Levuur, Mobiel21, Glassroots, ... that do wonderful projects in this field.

[175] Wallström, M. (2004). Reclaiming cuty streets for people. Chaos or quality of life? European Commission. Retrieved from http://ec.europa.eu/environment/pubs/pdf/streets_people.pdf

[176] Heath, C.; Heath, D. Switch: *How to Change Things When Change Is Hard;* Bantam Doubleday Dell Publishing Group: New York, NY, USA, 2010.

[177] Gifford, R.J. The dragons of inaction: Psychological barriers that limit climate change mitigation and adaptation. *Am. Psychol. 2011, 66,* 290-302.

[178] European Commission. Eurobarometer. https://ec.europa.eu/clima/citizens/citizen-support-climate-action_en#previous-surveys

[179] For example, web platforms Bol.com and Zalando offer an ever-increasing range of sustainable products, and also introduced ambitions and actions to make their (logistics) operations thoroughly sustainable. In its own country, Colruyt Group published a sustainability report with a selection of their efforts.[15] The retail group reports a reduction of 7.4% in emissions compared to the base year 2008, amongst others by implementing solutions for cooling, heating, transport and logistics.

[180] Kerret, D.; Orkibi, H.; Ronen, T. Testing a model linking environmental hope and self-control with students' positive emotions and environmental behaviour. *J. Environ. Educ. 2016, 47,* 307-317

[181] https://www.aviation24.be/railways/thalys-traffic-is-gradually-returning-to-normal-after-third-serious-incident-in-a-few-days/

[182] Macharis, C. and D. Kerret, 2019, "The 5E Model of Environmental Engagement: Bringing Sustainability Change to Higher Education through Positive Psychology", *Sustainability,* 11, 241, 13 pages, https://doi.org/10.3390/su11010241

[183] This new 6E model can be seen as an operationalisation of Dorit Kerret's positive sustainability model. In fact, it has much in common with Fran Bambust's 7E model for social marketing, I noticed afterwards. However, this is mainly aimed at communication campaigns by governments.

ENDNOTES

FACTOR 8

Colophon

Text
Cathy Macharis
The author has used the English translation of
Met een factor 8 naar de mobiliteit van de toekomst
to update her book by adding several new elements.

Illustrations
Mathilde Guegan

Translation
Ganiyat Temidayo Saliu

Layout
www.groupvandamme.eu

Published by
Stichting Kunstboek bv
Legeweg 165, 8020 Oostkamp (BE)
Tel. +32 50 46 19 10
info@stichtingkunstboek.com
www.stichtingkunstboek.com

ISBN 978-90-5856-701-7
D/2022/6407/20
NUR 740

Printed in the EU

FACTOR